MATHEMATICIANS AND THEIR GODS

Interactions between mathematics and religious beliefs

Edited by
SNEZANA LAWRENCE
and
MARK McCARTNEY

OXFORD
UNIVERSITY PRESS

OXFORD
UNIVERSITY PRESS

Great Clarendon Street, Oxford, OX2 6DP,
United Kingdom

Oxford University Press is a department of the University of Oxford.
It furthers the University's objective of excellence in research, scholarship,
and education by publishing worldwide. Oxford is a registered trade mark of
Oxford University Press in the UK and in certain other countries

First Edition published in 2015

Published in the United States of America by Oxford University Press
198 Madison Avenue, New York, NY 10016, United States of America

British Library Cataloguing in Publication Data
Data available

Library of Congress Control Number: 2014956601

ISBN 978–0–19–870305–1

PREFACE

The overlap between mathematics and theism may seem at first sight an unpromising topic for a book. But as the many contributors to this volume show, the intersection between the two provides fertile ground both for historians of mathematics and of theology, and for the interested general reader. The chapters have been written by a range of subject experts, who have given willingly of their time and their knowledge. They have also submitted with good humour and patience to our various editorial queries, suggestions, and proddings, and we thank them for their enthusiasm for, and commitment to, the project.

The chapters of the book are ordered broadly chronologically and thus the work can be read beginning with the Pythagoreans and then meandering through a range of times, places, and people, to end with Kurt Gödel. However, each chapter is independent of the others and so the reader can feel free to 'dip in' to the book at whichever point he or she finds most interesting. Hopefully, such a reader will then be enticed into the other chapters.

As is noted in the introductory chapter, this book makes no attempt to be comprehensive. The most glaring omission is of Islamic mathematics, with the centre of gravity of the book being within the Christian West. The bias is benign and can be accounted for by the interests of the editors, and the desire to keep the scale of the book within manageable proportions. If there is a benefit of this bias, it is that it illustrates that even within such a restricted field of study the interactions between religion and mathematics are surprisingly rich.

Snezana Lawrence & Mark McCartney
March 2015

CONTENTS

CHAPTER 1

Introduction

MARK McCARTNEY

It is unfortunate that in popular thought the prevailing description of the interaction between science and religion is one of warfare. God, in some circles it seems, is the enemy of progress, the person we lazily appeal to in order to fill the gaps in our current scientific knowledge, an outdated idea, a delusion, an intellectual non-subject. At best, religious belief is seen as a matter of private devotion which has nothing to do with public life. At worst, some quarters seem to want to take what may be called an Orwellian 'Animal Farm' approach to the matter, replacing Snowball the pig's 'four legs good, two legs bad', with a version of scientism which states 'science good, religion bad'.

And yet, dig down a little below the surface of popular assessment to scholarship on the history and philosophy of science, and a very different picture begins to appear. In academic circles the nineteenth-century 'conflict model' to describe the interaction between science and religion has been rejected and replaced with what Andrew Gregory in Chapter 2 of this book describes as the 'complexity model'. The complexity model recognizes that there is a spectrum of interactions, ranging from conflict or mutual non-engagement to dialogue and synthesis. It recognizes that there are indeed conflicts, but there are also connections and consonances, and that these interactions have cultural and historical contexts. Academics such as John Polkinghorne (1930–) and Alister McGrath (1953–) are examples of two scholars trained both in science and theology who have written widely on the links between those disciplines from a Christian perspective.[1] But they are only two from a long list of scholars, some of faith, some of none, who feel that there is much more to be said, and that much more needs to be said, than the Orwellian 'science good, religion bad'.[2]

However, move from science to mathematics and the suggestion that there is an overlap with theism is likely to lead to puzzled looks. Linking algebra with the Almighty seems as sensible as trying to find common ground between botany and basketball. The famous story of Leonhard Euler (1707–83) confronting Denis Diderot (1713–84) at the court of Catherine the Great springs to mind: Diderot arrived at the Russian court in the 1770s and, it is said, spoke freely of his atheism. While Catherine was amused, her advisors suggested Diderot should have his atheistic wings clipped. Hence Euler, then at the St Petersburg Academy, was brought before the court and publically confronted Diderot with the statement 'Sir, $(a + b^n)/n = x$; hence God exists – Respond!' Diderot, unable to answer, and humiliated by the laughter from all sides, left the court and immediately returned to France.

The tale is, of course, apocryphal.[3] It is amusing to imagine that the brilliant mathematician Euler would say such a thing, even more amusing that Diderot, editor of the *Encyclopédie* and well versed in mathematics, should have been taken in by it, and positively ridiculous that any such 'argument' could make an impression on anyone. We may seek for links between God and the scientific description of the created order, but surely any connections we might wish to make between God and our equations will be as weak as Euler's fictitious remark?

However, it is precisely with the phrases 'scientific description' and '*our* equations' that potential links between mathematics and theism are to be found.

The scientific description of nature

Galileo Galilei (1564–1642), in his book *The Assayer*, famously stated that 'Philosophy is written in this grand book – I mean the universe – which stands continually open to our gaze, but it cannot be understood unless one first learns to comprehend the language and interpret the characters in which it is written. It is written in the language of mathematics, and its characters are triangles, circles, and other geometrical figures, without which it is humanly impossible to understand a single word of it; without these one is wandering about in a dark labyrinth.'[4] Galileo's insight that mathematics is the language of the universe has grown in significance since he penned it in 1623. It is not simply that mathematics is the language of nature, but also that, as the Nobel laureate

Eugene Wigner (1902–95) pointed out in a famous essay, there is an 'unreasonable effectiveness of mathematics in the natural sciences'.[5] In his essay he notes that in theoretical physics, complex and independently discovered branches of mathematics often turn out to be the appropriate language of nature. Thus, in quantum mechanics the use of complex numbers is 'far from natural or simple and they cannot be suggested by physical observations. Furthermore, the use of complex numbers in this case is not a calculational trick of applied mathematics but comes close to being a necessity in the formulation of the laws of quantum mechanics.'[6]

He goes on to quip that:

> A possible explanation of the physicist's use of mathematics to formulate his laws of nature is that he is a somewhat irresponsible person. As a result, when he finds a connection between two quantities which resembles a connection well-known from mathematics, he will jump at the conclusion that the connection *is* that discussed in mathematics simply because he does not know of any other similar connection. It is not the intention of the present discussion to refute the charge that the physicist is a somewhat irresponsible person. Perhaps he is. However, it is important to point out that the mathematical formulation of the physicist's often crude experience leads in an uncanny number of cases to an amazingly accurate description of a large class of phenomena. This shows that the mathematical language has more to commend it than being the only language we can speak; it shows that it is, in a very real sense, the correct language.

Closely linked to the effectiveness of mathematics in the physical sciences is what many mathematicians see as the intrinsic beauty of good quality mathematics. Thus G.H. Hardy (1877–1947), one of the greatest British pure mathematicians of the twentieth century, stated that 'The mathematician's patterns, like the painter's or the poet's, must be beautiful; the ideas, like the colours or the words, must fit together in a harmonious way. Beauty is the first test; there is no permanent place in the world for ugly mathematics.'[7]

When asked during a lecture at Moscow University in 1957 to give his philosophy of physics, the Nobel prize winner Paul Dirac (1902–84) wrote, in capitals, on the blackboard 'PHYSICAL LAWS SHOULD HAVE MATHEMATICAL BEAUTY'.[8]

Anyone who has dealt with Maxwell's equations of electromagnetism, or the Schrödinger equation, cannot fail to be impressed with the concise elegance of the formulae. Their compact beauty is made all the more stark when one considers the very broad range of phenomena they ultimately explain and encapsulate.

Finally, it is hard to resist physicist George Gamow's (1904–68) typically witty assessment of the links between theory and experiment in physics which he put in a letter to Dirac in 1965:

Case I Trivial statement

If an elegant theory agrees with experiment, there is nothing to worry about.

Case II Heisenberg's postulate

If an elegant theory does not agree with experiment, the experiment must be wrong.

Case III Bohr's amendment

If an inelegant theory disagrees with experiment, the case is not lost because [by] improving the theory one can make it agree with experiment.

Case IV My opinion

If an inelegant theory agrees with experiment, the case is hopeless.[9]

Gamow's remark, for all its humour, has a grain of truth in it. Case IV is quite close to what many theoretical physicists believe.

However, what does mathematical beauty, or the effectiveness of mathematics in the physical sciences, have to do with God? In his essay Wigner uses words and phrases that could be read as having religious overtones. Thus, for example, he uses the word 'miracle' 12 times and states that the effectiveness of mathematics 'is something bordering on the mysterious' and is 'a wonderful gift which we neither understand nor deserve'.[10] That said, God is not mentioned and no religious conclusions are drawn. Indeed there are those who simply disagree with Wigner that there is *any* case to answer. They claim that the effectiveness of mathematics is quite reasonable, is not mysterious, and can be explained historically.[11] If mathematics is defined as the means to describe all possible patterns, then it is hardly surprising if a pattern-filled universe is described mathematically.

Further, even though Hardy, Dirac, and others see beauty as a key discriminator, perhaps even a guide to truth, in good mathematics or theoretical physics, they were not theists (Figure 1.1). Hardy took his atheism so seriously that he refused to enter college chapel even for formal business (though the amusing irony that there is a memorial plaque to him in the ante-chapel of Trinity College, Cambridge would not have been lost on his sense of humour). Dirac similarly was an atheist. In his twenties he directly stated 'If we are honest – and scientists have to be – we must admit that religion is a jumble of false assertions with no basis in reality. The very idea of God is a product of the human imagination.'[12] That said, in later life he made use of the Deity as a rhetorical device:

'God', he wrote in a *Scientific American* article in 1963, 'is a mathematician of a very high order, and He used very advanced mathematics in constructing the universe'.[13] It seems that when faced with the sheer grandeur of the universe even some atheists reach for the idea of a God they do not believe in. Steven Weinberg (1933–), who won the Nobel Prize for physics in 1979 and has written an eloquent if almost melancholic affirmation of atheism in his book *Dreams of a Final Theory*,[14] states that 'Whatever one's religion or lack of it, it is an irresistible metaphor to speak of the final laws of nature in terms of the mind of God.'[15]

There are those, however, who feel that it is more than a metaphor, and that there is a substantive link between theism and the ideas of beauty and effectiveness in mathematics. Thus John Polkinghorne states 'The "unreasonable effectiveness of mathematics" in uncovering the beauty of the physical world . . . is a hint of the presence of the Creator, given to us creatures who are made in the divine image. I do not present this conclusion as a logical demonstration – we are in the realm of metaphysical discourse where such certainty is not available to either believer or to unbeliever – but I do present it as a coherent and intellectually satisfying understanding.'[16] Elsewhere Polkinghorne states 'The physical universe seems shot through with signs of mind. That is indeed so, says the theist, for it is God's Mind that lies behind its rational beauty.'[17]

As Polkinghorne says, these ideas are not in any sense proofs for God's existence, and indeed no theist would suggest that they are. Rather, for the theist, the ideas of beauty and effectiveness in mathematics sit with a suggestive coherence within their worldview.

Whose equations are they?

Linked with the idea of the effectiveness of mathematics is the issue of what is called mathematical Platonism. Simply put, mathematical Platonism states that when doing mathematics we are *discovering* rather than inventing results, and thus mathematics is, in some sense, independent of us.

A well-known contemporary exponent of mathematical Platonism is the Oxford mathematical physicist Roger Penrose (1931–) who states 'I have made no secret of the fact that my sympathies lie strongly with the Platonistic view that mathematical truth is absolute, external and eternal, and not based on man-made criteria; and that mathematical objects have a timeless existence of their own, not dependent on human society nor on particular physical objects.'[18]

(a)

Figure 1.1 The diversity of the twentieth century mathematical mind: (a) G.H. Hardy, (b) Paul Dirac, and (c) John Polkinghorne. All three of these men were professors at Cambridge during the twentieth century, Hardy in pure mathematics, and Dirac and Polkinghorne in theoretical physics. Between them they reveal aspects of the diversity of views on religion within the mathematical sciences. Dirac, an atheist, saw religion as a 'jumble of false assertions', but also used the metaphor of God as the mathematician behind the universe. Hardy was also an atheist, but with sharp wit also declared that God was his 'personal enemy'. When going to watch cricket he would take an 'anti-God battery' of sweaters, work, and an umbrella to fool the Almighty into thinking that he was expecting poor weather. Hardy argued that God would then send sunshine, which was, of course, what Hardy had really wanted all along. Polkinghorne, a Christian, resigned from his chair in mathematical physics aged 50 and retrained as an Anglican Priest. He has written extensively on the interactions between science and religion. (Image of Dirac courtesy of Florida State University, taken while Dirac was in the Faculty at FSU. Image of Polkinghorne from Wikimedia Commons.)

(b)

(c)

Figure 1.1 *Continued*

Similarly G.H. Hardy states, 'I will state my own position dogmatically in order to avoid minor misapprehensions. I believe that mathematical reality lies outside us, that our function is to discover or *observe* it, and that the theorems which we prove and which we describe grandiloquently as our "creations", are simply our notes of our observations.'[19] Later in the same book Hardy provocatively states:

At any rate . . . [the] realistic view is much more plausible of mathematical than of physical reality, because mathematical objects are so much more what they seem. A chair or a star is not in the least like what it seems to be; the more we think of it, the fuzzier its outlines become in the haze of sensation which surrounds it; but "2" or "317" has nothing to do with sensation, and its properties stand out the more clearly the more closely we scrutinize it . . . 317 is a prime, not because we think so, or because our minds are shaped in one way rather than another, but *because it is so*, because mathematical reality is built that way.[20]

Finally, Paul Erdös (1913–96), the Hungarian pure mathematician who was eccentric, brilliant, and prolific in equal measure, remained agnostic about the existence of the 'SF' or 'Supreme Fascist', as he described the Deity, but nevertheless he famously believed in the existence of 'The Book': 'I'm not qualified to say whether or not God exists. I kind of doubt he does. Nevertheless, I'm always saying that the SF has this transfinite Book – transfinite being a concept in mathematics that is larger than infinite – that contains the best proofs of all mathematical theorems, proofs that are elegant and perfect.'[21] Though not often expressed as directly as by Penrose and Hardy, there is a strong sense among many mathematicians that when going about their research they are indeed *discovering* new truths. However, if pressed, they would probably deny they are mathematical Platonists – nervous, perhaps, of suggesting belief in a separately existing mathematical realm.

Mathematical Platonism has a long history[22] with many adherents through the ages and it is not surprising that it has been linked with theism. Within Christianity the natural move of placing mathematical truths within the mind of God goes back as far as Augustine. However, in the absence of God the very idea of, to quote Penrose again, an 'absolute, external and eternal' mathematical world, is likely to strike some as at worst a Baroque metaphysical assertion, or at best the mistaking of the uncovering of logical consequence for the discovery of independent truth.[23]

As the American philosopher Alvin Plantinga (1932–) states:

most people who have thought about [mathematical Platonism] think it incredible that these abstract objects should just exist, just be there, whether or not they are thought of by anyone, more broadly, whether or not they are the object of any kind of mental or intellectual activity . . . It is therefore extremely tempting to think of abstract objects as ontologically dependent upon mental or intellectual activity in such a way that either they are just thoughts, or at least at any rate they couldn't exist if not thought of . . . On the other hand, if abstract objects were divine thoughts, there would be no problem here. So perhaps the most natural way to think about abstract objects, including numbers, is as divine thoughts.[24]

As with effectiveness and beauty, it is important to be clear that no theist will see mathematical Platonism as a proof for, or necessarily leading to, theism. Rather they find a satisfying coherence between the two ideas. Further, it should be noted that not all mathematical Platonists are theists (Penrose and Hardy are certainly not) and that not all mathematicians who are theists are mathematical Platonists. As a counter-example, the Lutheran and eminent Stanford computer scientist Donald Knuth (1938–) states 'mathematics and computer science are the two *unnatural* sciences – the study of things that are created by people instead of being present in nature . . . In mathematics and computer science, we can actually prove theorems, solve problems and know that we have an answer, because we get to make up the ground rules.'[25]

Mathematicians and their gods

Though ideas of mathematical effectiveness, beauty, and Platonism are all invoked in modern discussions about mathematics and theism, a defence or historical survey of the possible links between these ideas is *not* the purpose of this book (though they will occasionally recur within its chapters). Rather the aim is much more practical and interesting. To understand fully any individual figure, or movement, from history we must take into account the surrounding cultural, social, and philosophical influences. We should do this not simply because it is good scholarship and fleshes out the historical context, but also because it can lead to fascinating insights which may otherwise be overlooked, or may even provide the key to a proper understanding of some aspect of a topic. If this is true of cultural, social, and philosophical influences, it is also true of theological ones. Many scientists and mathematicians, from antiquity to the present, have dwelt on the nature of God and on God's relationship with the world, themselves, and their studies. Others, while perhaps having no deep concerns about divinity, have lived in times that were deeply religious. In some cases these connections are essential to enable a proper understanding of the person's life, in others they allow new insights and fresh perspectives. This book brings together essays that highlight examples of these fresh perspectives across the history of the mathematical sciences, a term we use in its broad sense, and hence in what follows we include topics such as medieval optics, Renaissance astronomy, and nineteenth-century natural philosophy.

It will be noted that the material covered in this book is biased towards Western Christendom. The most significant omission is that we have ignored the rich vein of Islamic mathematics that runs throughout the medieval era. This reflects, at least in part, the interests of the editors, but it was also a self-imposed restriction to help keep the book within manageable proportions. However, even in this restricted field the potential material is substantial and we have necessarily been selective. There has been room for only a sample of the historical movements and characters where God and mathematicians intersect.

Even as I write, names of those not included such as John Napier (1550–1617), Gottfried von Leibniz (1646–1716), Leonhard Euler (1707–83), Charles Babbage (1791–1871), G.G. Stokes (1819–1903), and E.T. Whittaker (1873–1965) all clamour for attention. Napier wrote a commentary on the New Testament Book of Revelation in 1593. It was a book that went through many editions, and was translated into French, German, and Dutch. In Napier's view it was his most important work (Figure 1.2). Leibniz produced a solution to the problem of evil where the world in which we find ourselves is regarded as the best of all possible worlds (a view mercilessly caricatured in Voltaire's *Candide*). Euler, easily one of the greatest mathematicians of all time wrote, in addition to his prodigious mathematical output, a *Defence of the Divine Revelation against the Objections of the Freethinkers* (1747). Babbage, 'father of the computer' and Lucasian professor of mathematics at Cambridge wrote *The Ninth Bridgewater Treatise: A Fragment* (1837). It covered a wide range of topics, including free will and the nature of miracles. The book is entitled *A Fragment* because some chapters are incomplete – most notably the chapter on the origin of evil, which is entirely blank save for a one sentence note to explain its absence (Figure 1.3). G.G. Stokes, who also held the Lucasian chair of mathematics at Cambridge, gave the Gifford lectures (a prestigious endowed lecture series in natural theology held in the ancient Scottish universities) and wrote a book defending conditional immortality, that is the idea that the human soul is immortal only if salvation through Christ is accepted. Finally, E.T. Whittaker, professor of mathematics at Edinburgh, wrote extensively defending the Christian faith, including *Space and Spirit: Theories of the Universe and Arguments for the Existence of God* (1946).

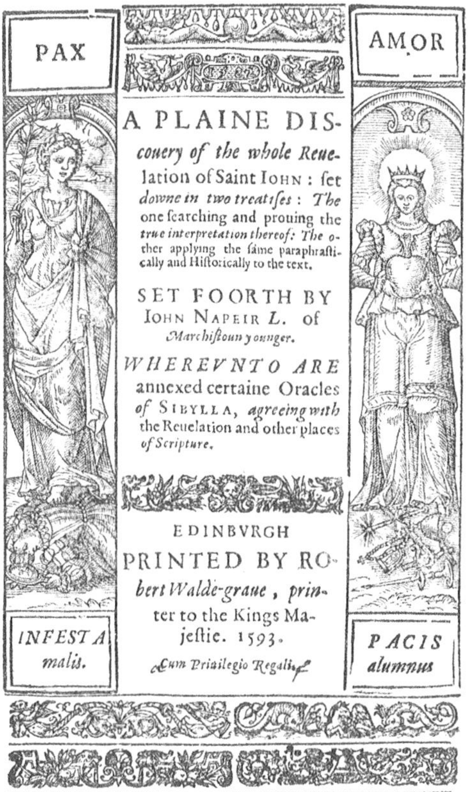

PAX

AMOR

A PLAINE DIS-
couery of the whole Reue-
lation of Saint IOHN : ſet
downe in two treatiſes : The
one ſearching and prouing the
true interpretation thereof: The o-
ther applying the ſame paraphraſti-
cally and Hiſtorically to the text.

SET FOORTH BY
IOHN NAPEIR L. of
Marchiſtoun younger.

WHEREVNTO ARE
annexed certaine Oracles
of SIBYLLA, agreeing with
the Reuelation and other places
of Scripture.

EDINBVRGH
PRINTED BY RO-
bert Walde-graue, prin-
ter to the Kings Ma-
jeſtie. 1593.
Cum Priuilegio Regali.

INFESTA
malis.

PACIS
alumnus

Figure 1.2 Title page of John Napier's 1593 *A Plaine Discovery, of the whole Revelation of Saint John* in which he concluded that the Pope was the Antichrist, and that the Day of Judgement would occur between 1688 and 1700.

But they will all have to wait for another book. However, perhaps the selection given in the chapters that follow will enthuse the reader to study some of those names for themselves.

A bird's eye view of the rest of this book

Each chapter of this book is self-contained and can be read independently of the others. However, to aid historical navigation, they are arranged roughly chronologically. We begin with Pythagoras in Chapter 2. Alas, as Andrew Gregory points out, we have no direct evidence that he was the author of the eponymous theorem, and Pythagoras emerges as a shadowy figure from ancient history, as much a ritualist and wonder worker as a mathematician. However, about Pythagoras'

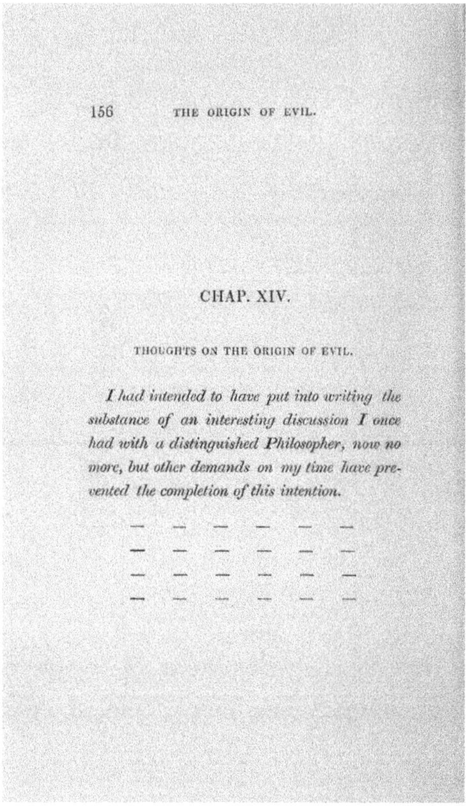
156 THE ORIGIN OF EVIL.

CHAP. XIV.

THOUGHTS ON THE ORIGIN OF EVIL.

I had intended to have put into writing the substance of an interesting discussion I once had with a distinguished Philosopher, now no more, but other demands on my time have prevented the completion of this intention.

Figure 1.3 Chapter 14 of Charles Babbage's *Ninth Bridgewater Treatise* in its entirety. The problem of evil can be stated simply as follows: If God is both good and all-powerful, why is there so much evil and suffering in the world? Although simple to state, this dilemma is considered by many to be one of the most serious objections to theism, and much ink has been spilt in discussing it. Babbage provides what must be the shortest and least helpful discussion of the matter ever given in print.

followers, the Pythagoreans, much more can be said. They attributed symbolic meaning to some numbers, and Aristotle stated that 'They construct the whole heaven out of numbers', though what exactly that meant is unclear. Their idea that the cosmos sang with the music of the celestial spheres is one which resonated down the millennia, still being heard by Dante in the *Paradiso* and by Dylan Thomas in *Under Milk Wood*. Of their view of the gods, we know very little. They do not appear to have believed in a creator god, and their spiritual thought centred on how the soul migrated to other living things after death, rather than worship of the divine. Gregory gives a wide ranging account of Pythagorean thought on mathematics, music, and cosmology, which apart from being fascinating in its own right gives helpful background for the chapters by Snezana Lawrence on Freemasonry (Chapter 9) and Jean-Pierre Brach on mystical arithmetic in the Renaissance (Chapter 6).

Allan Chapman (Chapter 3) traces aspects of optical science and mathematics through more than two millennia, from the Greeks, through medieval Islamic and Christian thought, to Hooke and Newton. Whereas for the Pythagoreans the gods and their role in creation seems to have been of little interest, and for Platonic thought the demiurge (literally the public worker, or craftsman) constructed the world from pre-existing material, with Christian and Islamic thought we have God as creator, and light itself as an image of His majesty. Chapman sees the work of Muslims such as Al-Kindi in the ninth century and Alhazan in the tenth–eleventh, or the thirteenth-century work of the Franciscan Roger Bacon and the Dominican Theodoric of Freiburg, as being undergirded by the idea of God. Belief in God, and a desire to know more of His creation and His nature were, Chapman argues, at least partial motivators for scientific discovery for these men.

Johannes Kepler (1571–1630) studied theology at the University of Tübingen with the aim of becoming a Lutheran pastor, but instead he became diverted into astronomy. As Owen Gingerich shows in Chapter 4, his theology was an important aspect of his science. In the geocentric model of a cosmos of heavenly spheres, the outermost sphere, beyond the fixed stars, is the *primum mobile* or first movable. Beyond that is God Himself. God, the Unmoved Mover, gives motion to the *primum mobile* not by some sort of positive action – for God is unmoved – but rather the *primum mobile* is moved by its love for God, and thus being in motion transfers that motion to the inner spheres. Such a cosmology may appear contrived, or even quaint, to us in the twenty-first century, but in Kepler's age it was still embedded solidly in culture and for centuries had been part of a coherent, all-encompassing worldview.[26] Kepler kept at least one part of the old cosmos – God as the source of motion – in his heliocentric universe. Placing the sun at the centre of the cosmos, the closer a planet was to the sun, the faster it moved. Next, taking the sphere as an image of the Trinity with the centre representing God the Father, the surface God the Son, and the intermediate space God the Holy Spirit, it was straightforward to identify the centre – God the Father – as the prime mover of the planets. Gingerich's statement that 'This Trinitarian image of the solar system must have been one of the strongest reasons for Kepler's accepting a sun-centred cosmos' highlights an example of the surprising interplay between theism and science throughout history.

Robin Wilson and the late John Fauvel[27] give a fascinating overview of the connections between religious thought and combinatorics in the Renaissance in Chapter 5. Behind what the modern reader might dismiss as exercises worthy only to set a student – finding the number of possible arrangements of letters in the name Jesus, or of the words of a line of Latin praise to the Virgin – are much grander aspirations. Scholars such as Ramon Lull (*c.* 1232–1316), and following him Athanasius Kircher (1601–80), aimed at a complete and unified system of knowledge with combinatorics providing, as Wilson and Fauvel state, 'the basic tool for exploring all that can be known'. The Greek inscription that forms part of the frontispiece of Kircher's 1669 book *Ars Magna Sciendi Sive Combinatoria* reads 'Nothing is more beautiful than to know the All'. For Kircher that *All* encompassed science, art, and divinity, with the mathematics of combinatorics providing a key tool of interrogation. The literary scholar and author C.S. Lewis (1898–1963) stated that 'it is more important that Heaven should exist than that any of us should reach it'.[28] The equivalent scientific statement is perhaps 'it is more important that an ultimate unity of knowledge should exist than that any of us should find it'; however, in the seventeenth century of Kircher the boundaries of that knowledge were wider and more diverse than those admissible by modern science.

Across the Early Modern period (fifteenth–seventeenth centuries) mathematics was used to help unify knowledge and explain the universe. Jean-Pierre Brach shows in Chapter 6 how the Pythagoreanism described by Andrew Gregory was able to have a new life at the beginning of modernity, with numbers having an ontological existence and possessing symbolic meanings which, although they may seem strange to us, made sense in their time. Great Humanist minds used number symbolism as a common thread between cosmology, magic, Christian theology, and Christian scripture. It is interesting that a common link between Jean-Pierre Brach's and Snezana Lawrence's chapters is the survival of aspects of Pythagorean mysticism in Humanism and Freemasonry respectively. That said, by the end of the seventeenth century number had lost virtually all of its qualitative significance for the scholar. Any mystical symbolic links of three with the Trinity, or twelve with the spread of the Trinitarian Gospel to all four corners of the earth, or the square root of 666 with the evils of Catholicism (as described in Rob Iliffe's Chapter 7), was evaporating. Numbers now meant no more than themselves, and the Platonic solids no longer had the cosmological significance that Kepler gave them when he embedded the planetary spheres between their surfaces (see Figure 4.1).

It is perhaps not surprising that Newton (1642–1727) believed that the *imago dei* – that aspect of humanity which sets us apart as being created in God's image

(Genesis 1 v 27) – should be found in a human's ability to reason.[29] God was, after all, the ultimate reasoner, the only possible being capable of calculating the subtle variations in motion resulting from the multiple gravitational interactions in the solar system. But what is less well known is that Newton's famously complex, prickly, and obsessive character extended to his religious beliefs. His poring over the Bible and church history led him to view the orthodox Christian view of God as Trinity as in fact a heresy, though he was careful to keep his views on the matter private. He made detailed studies of the apocalyptic books of Daniel in the Old Testament, and of Revelation in the New, and while those too he kept broadly to himself, Rob Iliffe shows in his Chapter 7 that such interest in prophesy and, ultimately, the second coming of Christ was not unusual at the time. Neither was Newton the only mathematician who was intrigued. As noted earlier John Napier, of logarithm fame, published *A Plaine Discovery, of the whole Revelation of Saint John* in 1593 (Figure 1.2).

On the other side of Europe from an elderly Newton in London, Maria Gaetana Agnesi was born in Milan in 1718. Though an enthusiast for Newton's fluxions, in other ways the two could hardly have been more different. Newton was fiercely anti-Catholic, could not cope with public criticism, and cared relatively little for his fellow man. In contrast, Agnesi was a devout Catholic who, as a child prodigy, found herself involved in public debates that were staged by her father to further his social position, and when she grew up she devoted herself to teaching the young and looking after the poor. Whereas Newton lay in a tradition that saw the wonder of creation as a pointer to God, Agnesi saw mathematics as a discipline for the mind, which could be part of a spiritual exercise that aided the more serious business of mediation on the Divine nature. Massimo Mazzotti in Chapter 8 shows that Agnesi is a case study in the need to treat each historical figure on their own terms and as an integrated whole. As he states 'the very categories of "science" and "religion" as referring to two incompatible sets of practices would be meaningless to Agnesi'. She inhabited a mental world where no such dichotomy existed.

Returning to London, the origin of the Masonic order at the beginning of the eighteenth century, its self-created mythology involving Euclid, Pythagoras' theorem, and King Solomon, and its social, moral, and mystical aims are discussed by Snezana Lawrence in Chapter 9. Freemasonry, detached from any traditional theistic beliefs, provides a frame for a fascinating window onto aspects of the eighteenth and nineteenth century. Starting from practical guilds for builders, Lawrence takes us on a tour encompassing amongst other things, strange Masonic initiation rites, Gaspard Monge's descriptive geometry, and the Rosetta Stone. Lawrence suggests that much of what might be described as 'sacred geometry' in the popular imagination, with its hidden knowledge and conspiratorial edges, finds root in the Freemasonry of a previous age. As she wryly observes, Freemasonry 'created the concept of "sacred geometry" for better or for worse, worse probably being the view of the majority of mathematicians in modern times'.

As we enter the nineteenth century, Mark Richards in Chapter 10 shows that there was much more to Lewis Carroll than Alice. To understand the man fully we need to take into account not only his life as a mathematician, but also the importance of his Christian faith. Indeed as Richards notes, Carroll saw his work in mathematics as 'Work for God'. Elizabeth Lewis picks up on this idea in her Chapter 11 on the book *The Unseen Universe*, by two other Victorians, Tait and Stewart, when she notes that their writing of the book 'might even be considered to constitute religious service'. However, while Carroll saw his life as a teacher and mathematical communicator as a quiet vocation, Tait and Stewart saw their book as an overt apologetic and evangelical work showing the unity of the Christian faith with the latest ideas in Victorian natural philosophy – conservation of energy, the heat death of the universe, and the vortex atom. The book, largely forgotten now, was controversial amongst Christian and agnostic alike, and had the added fascination of initially being published anonymously. Lewis ably resurrects the book, the authors, and the controversy, for the modern reader. *The Unseen Universe* was a remarkable attempt at providing a grand unified theory of science and faith, but if it has an abiding lesson for the modern reader it is, perhaps, the danger of seeking detailed linkages between the latest science and theology.

A book which, at least in the mathematics community, needs no such resurrection is discussed by Mel Bayley, who provides us with a fascinating perspective on E.A. Abbott's enduring classic *Flatland* in Chapter 12. Though the book is familiar to many modern readers, Bayley shows, by setting the text within the context of Abbott's other works on theological matters and within the wider context of nineteenth-century mathematics, that *Flatland* is much more than delightful Victorian whimsy. Within *Flatland* is a world of theological observation and critique. Bayley reads it as 'a cautionary tale about Victorian dogma and doctrine', with Abbott seeing the rise of higher dimensional geometry as pointing towards the need to embrace a more liberal, less miraculous, view of the Christian faith. It should be noted that both Abbott and Tait and Stewart sought to accommodate Darwinian evolution into their theistic worldview. Surprisingly though, the theologically more liberal (in a nineteenth-century context)

Abbott saw not God, but Satan as the force behind natural selection: 'we must learn to think, not of "Evolution by itself," but of "Evolution with Satan"'. Accepting Darwin as the devil's work is hardly a ringing theological endorsement of evolution by natural selection, but it highlights the fact that there were a wide range of religious responses to Darwin. Some rejected Darwinism as incompatible with Christianity, some accepted it warily, but others absorbed it into the Christian worldview with relatively little trouble.[30]

Finally, with Tony Anderson's Chapter 13, we arrive in the twentieth century and perhaps mathematic's greatest logician, Kurt Gödel. In February 1970 Gödel showed Dana Scott, then professor of philosophy and mathematics at Princeton, a manuscript containing his ontological proof for the existence of God. Gödel's thoughts on this proof date back to as early as 1941, and though it is work he never published, it became widely known in academic circles because Scott used the proof in a seminar at Princeton during the autumn of 1970. It seems, however, that Gödel saw the work as an exercise in logic, rather than anything he himself was convinced by. Ontological arguments for God's existence date back to Anselm of Canterbury in the eleventh century, with famous mathematicians such as Descartes and Leibniz also providing versions. The status of such arguments within theology is mixed, with some finding them useful, some dismissing them as inadequate, and others finding them simply inappropriate; mistaking God as someone we, as mere creatures, can pronounce on. However, as the chapters of this book show, few mathematicians have attempted to make such direct use of their art upon divinity. Rather, they lived rich and fascinating lives in which their mathematical work, and religious thought and action, formed part of a complex and entangled whole.

Notes and references

1. See, for example, A.E. McGrath, *Science and Religion: A New Introduction*, 2nd edn, Wiley-Blackwell, 2010, or J. Polkinghorne, *Science and Religion in Quest of Truth*, Yale University Press, 2011.
2. The literature here is vast. For the interaction between science and religion in historical perspective see J.H. Brooke, *Science and Religion: Some Historical Perspectives*, Cambridge University Press, 1991, or R.L. Numbers (ed.), *Galileo Goes to Jail, and Other Myths about Science and Religion*, Harvard University Press, 2009. For a more general introduction see McGrath, *Science and Religion*. For more personal perspectives from respectively a practising geneticist, physicist, and computer scientist see F. Collins, *The Language of God*, Free Press, 2006; O. Gingerich, *God's Universe*,

Belknap Press, 2006; D.E. Knuth, *Things a Computer Scientist Rarely Talks About*, CSLI Publications, 2001.

3. B.H. Brown, The Euler-Diderot Anecdote, *American Mathematical Monthly*, **49**, 1942, 302–3, reprinted in W. Dunham (ed.), *The Genius of Euler*, Mathematical Association of America, 2007, pp. 57–9.
4. Quoted in A. Frova and M. Marenzana, *Thus Spoke Galileo*, Oxford University Press, 2006, p. 402.
5. E.P. Wigner, The unreasonable effectiveness of Mathematics in the Natural Sciences, *Communications on Pure and Applied Mathematics*, **13**(1), 1960, 1–13. Reprinted in E.P. Wigner, *Symmetries and Reflections: Scientific Essays*, Indiana University Press, 1967, pp. 222–37.
6. E.P. Wigner, The unreasonable effectiveness of Mathematics, p. 229.
7. G.H. Hardy, *A Mathematician's Apology*, Canto, Cambridge University Press, 1992, p. 85.
8. G. Farmelo, *The Strangest Man: The Hidden Life of Paul Dirac, Quantum Genius*, Faber & Faber, 2009, p. 359.
9. G. Farmelo, *The Strangest Man*, p. 380.
10. E.P. Wigner, The Unreasonable Effectiveness of Mathematics, p. 223, p. 237.
11. I. Grattan-Guinness, Solving Wigner's mystery, *The Mathematical Intelligencer*, **30**(3), 2008, 7–17. J. Gray, The vagueness of Wigner's analysis, *Interdisciplinary Science Reviews*, **36**(3), 2011, 214–28. John D. Barrow is a well known expositor of the understanding of mathematics as the study of 'all possible patterns'.
12. W. Heisenberg, *Physics and Beyond: Encounters and Conversations*, Harper Collins, 1971, p. 85.
13. P.A.M. Dirac, The evolution of the physicist's picture of nature, *Scientific American*, **208**(5), 1963, 45–53.
14. S. Weinberg, *Dreams of a Final Theory*, Hutchison Radius, 1993, chapter 11.
15. S. Weinberg, *Dreams*, p. 194.
16. J. Polkinghorne, *Belief in God in an Age of Science*, Yale University Press, 1998, pp. 4–5.
17. J. Polkinghorne, *Reason and Reality: The Relationship between Science and Theology*, SPCK, 1991, pp. 76–7.
18. R. Penrose, *The Emperor's New Mind*, Oxford University Press, 1989, p. 151.
19. G.H. Hardy, *A Mathematician's Apology*, pp. 123–4.
20. G.H. Hardy, *A Mathematician's Apology*, p. 130.
21. P. Hoffman, *The Man Who Loved Only Numbers*, Fourth Estate, 1998, p. 26. The idea of Erdös's book of the most elegant proof's has since been brought to print in M. Aigner and G.M. Ziegler, *Proofs from THE BOOK*, 4th edn, Springer, 2010.
22. For a helpful overview, see D.J. Cohen, *Equations from God: Pure Mathematics and Victorian Faith*, Johns Hopkins University Press, 2007, chapter 1.

23. See for example, M. Leng, Creation and discovery in mathematics, in J. Polkinghorne (ed.), *Meaning in Mathematics*, Oxford University Press, 2011.

24. A. Plantinga, Theism and mathematics, *Theology and Science*, **6**(1), 2011, 27–33, p. 30.

25. D.E. Knuth, *Things a Computer Scientist Rarely Talks About*, CSLI Lecture Notes, 2001, pp. 217–18

26. For an indication of how complete and coherent that worldview was, see C.S. Lewis, *The Discarded Image*, Cambridge University Press, 1964. Describing the medieval worldview Lewis states in the epilogue of the book that 'Few constructions of the imagination seem to me to have combined splendour, sobriety, and coherence in the same degree.'

27. John Fauvel died in 2001. The chapter in this book is a revised and extended version of an article by Fauvel and Wilson originally published in the *Bulletin of the Institute of Combinatorics and its Applications* in 1994.

28. C.S. Lewis, *Surprised by Joy*, Geoffrey Bles, 1955, p. 199.

29. The location of the *imago dei* in Christian thought is typically placed either in humankind's ability to reason, to enter into relationship, or with humanity's regency over the created order. These 'three Rs' are not mutually exclusive. A helpful overview can be found in N.L. Herzfeld, *In Our Image: Artificial Intelligence and the Human Spirit*, Fortress Press, 2002, chapter 2.

30. For examples of the varieties of responses to Darwin see, for example, D.N. Livingstone, *Darwin's Forgotten Defenders: The Encounter between Evangelical Theology and Evolutionary Thought*, Regent College Publishing, 2001, or D.N. Livingstone, *Dealing with Darwin: Place, Politics, and Rhetoric in Religious Engagements with Evolution* (Medicine, Science, and Religion in Historical Context), Johns Hopkins University Press, 2014.

CHAPTER 2

The Pythagoreans: number and numerology

ANDREW GREGORY

There is a common perception of the ancient Greek thinker Pythagoras (*c.* 570–*c.* 495 BCE) as a mathematician and geometer, famed for his discovery of Pythagoras' theorem. Pythagoras has also been seen as a pioneer in the application of mathematics to music theory, as a champion of the importance of mathematics in understanding the cosmos, and as the originator of the idea of music of the heavenly spheres.

Modern scholarship over the last 40 years has developed a rather different picture of Pythagoras as someone interested in the fate of the soul after death, an expert on religious ritual, perhaps a shaman, a wonder worker, and founder of a religious sect. While the common perception of Pythagoras is certainly in need of some modification, these two pictures are not necessarily incompatible.

Older views of the relation of religion and science may have seen religion and science as incompatible or in some form of inevitable conflict with each other. Dating from the late nineteenth century, these views are known as the 'conflict' theory. More modern views recognize that science and religion can interact in many ways and that for many thinkers, there is compatibility between their scientific and religious views. While not denying there can at times be conflict between science and religion, the modern 'complexity' view also allows there to be other relations, such as support, symbiosis, compatibility or indifference depending on specific circumstances.

The major problem in developing an accurate picture of Pythagoras is that Pythagoras himself wrote nothing, and if his contemporaries wrote anything about him nothing of this has survived. It may be that those who associated with Pythagoras deliberately kept silent about his key views. What we know of Pythagoras comes from much later sources, many of which are unreliable.[1] There was an unfortunate tendency after Plato (428/427–348/347 BCE) and Aristotle (384–322 BCE) for Pythagoras to be built up as a semi-divine or a divinely inspired figure and visionary. Often the views of later thinkers were attributed to him, especially those of the later Pythagoreans, and Pythagoras was also credited with originating aspects of Plato's metaphysics and cosmology. The 'Pythagorean question' is that of the extent to which we can reconstruct the historical views of Pythagoras from the information we have available.[2]

The key turning point in modern studies of Pythagoras has been Walter Burkert's *Lore and Science in Ancient Pythagoreanism*.[3] Burkert analysed the available evidence and concluded that to find out about Pythagoras, we must look to the earliest and least corrupt sources, which essentially means looking at the evidence of Plato and Aristotle. It is this move in what we see as reliable evidence that has effected the shift away from the view of Pythagoras as the master mathematician towards Pythagoras as the religious leader.

One thing to emphasize early on is that Pythagoreanism was never a tight body of doctrine or a rigid system of beliefs. There was a great diversity of views among Pythagoreans on issues of religion, the nature of numbers and the application of numbers in our understanding of the cosmos. We know this from the fragments that have survived from thinkers such as Philolaus and Archytas and the reports of Plato and Aristotle.[4]

Pythagoras and the early Pythagoreans

There is very little that we can say for certain about Pythagoras. Pythagoras was born on the Greek island of Samos *c.* 570 BCE and died *c.* 490 BCE. Around 530 BCE he relocated to Croton in southern Italy, which became a centre for the Pythagoreans. It is said that Pythagoras travelled widely in his youth, to Egypt and other parts of Africa, to Babylonia and possibly even to India. When Pythagoras was considered an important mathematician it was speculated that he got at least some of his mathematical knowledge from his travels to Egypt. Nowadays his travels are seen as an attempt to gain knowledge of various esoteric religious cults in these places.

In what follows, I am going to begin by looking at Pythagoras and the various issues concerning mathematics and religious practice that relate to him. There will be little here about any specific god. The Pythagoreans were not monotheists and their religious practices centred around how to be pure in this life and how best to pass into the next life here on earth rather than religious worship. I will also look at some followers of Pythagoras who are important in the history of mathematics for various reasons – Hippasus, Philolaus, and Archytas – as well as two groups of Pythagorean followers – the *acousmatikoi* (the listeners) and the *mathematikoi* (the learners) – who had rather different attitudes to the Pythagorean tradition. Finally, I will look at Plato, who while not a Pythagorean himself, was clearly influenced by Pythagorean ideas. How Plato treats these ideas can also throw some light on explaining what the Pythagoreans may have been trying to do with those ideas.

Pythagoras' theorem

Did Pythagoras discover what we now know as Pythagoras' theorem? This now seems unlikely though it is possible that either Pythagoras or another early Pythagorean made some sort of contribution. It is not unusual to find discoveries or inventions credited to the ancient Greeks when they merely improved on something that had been invented or discovered earlier. The Archimedes screw, a device for raising water, was used much earlier by the Babylonians but was named after Archimedes who made significant improvements to the efficiency of the device.

Much depends here on exactly what we mean by 'discover' when we ask if Pythagoras discovered Pythagoras' theorem. Pythagorean triples, which are integer lengths for right angled triangles that conform to Pythagoras' theorem, were known a long time before Pythagoras and were well known to the Babylonians. The simplest example here is 3, 4, 5 where $3^2 + 4^2 = 5^2$; other examples are 5, 12, 13 and 8, 15, 17 and 7, 24, 25 (there were many more known in antiquity). The Babylonians though, as far as we are aware, did not have a general expression for Pythagoras' theorem nor did they have a proof of it. The Babylonians were in many ways excellent mathematicians but tended to restrict themselves to the practical application of mathematics rather than investigate the abstract or concern themselves with proofs. It is also unlikely that Pythagoras provided a proof of the theorem. If he did, we do not know the nature of the proof and it is quite early in the history of Greek mathematics for the concept of proof. There are other things that Pythagoras may have done such that his name became associated with the theorem. It is possible that he formulated the theorem in an abstract, general manner which perhaps had not been done before, perhaps he produced a significant diagram, or perhaps he simply celebrated someone else generating the proof. The often repeated story that Pythagoras sacrificed oxen on discovering the theorem does not look reliable, as the Pythagoreans were vegetarians and also believed that the human soul survived death and was reincarnated, possibly in humans, possibly in animals (see below).

Pythagoras is not given the credit for a proof of Pythagoras' theorem, nor seen as an important mathematician or geometer, by either Plato or Aristotle. Nor is Pythagoras seen as a significant contributor to mathematics or geometry by early histories of Greek mathematics.

It is significant that while both Plato and Aristotle talk of presocratic natural philosophy, they do not give Pythagoras any important role in this.[5] Plato, who says remarkably little about Pythagoras himself, says that:

> Such was Pythagoras, who was particularly beloved in this way, and his followers have a reputation for a way of life they call Pythagorean even down to this day.[6]

The picture of Pythagoras and the Pythagorean way of life that emerges from looking at the evidence in Plato and Aristotle is of someone whose key beliefs were in the immortality of the soul and reincarnation and whose expertise was in the fate of the soul after death and in the nature of religious ritual. Pythagoras' major achievements are seen as the advocacy and the founding of a way of life based on stringent dietary regulations, strict self-discipline, and the keen observance of religious ritual. Pythagoras, or perhaps the early Pythagoreans, may have contributed something to our understanding of right angled triangles, but it is unlikely that this is the outright discovery or proof of what we now know as Pythagoras' theorem.

Metempsychosis

The idea that the soul survives the death of the body and then can reincarnate, either in another human body or in an animal body, is known as metempsychosis.

We have reasonably solid evidence that this was indeed Pythagoras' view. Diogenes Laertius, an ancient doxographer, tells us that:

On the subject of reincarnation, Xenophanes tells a tale which begins: Now I turn to another account and I will show the way. He says this about Pythagoras: Once he passed a young dog which was being mistreated, and taking pity he said: 'Stop, do not beat it, that is the soul of a man who was my friend, I recognised it when it cried aloud.'[7]

There is, though, some consensus that this is a significant move away from the Homeric conception of the fate of the soul, which was rather bleak. The standard passage for comparison in Homer is where Achilles says:

I would rather be above ground still and labouring for some poor and portionless man, than be lord over all the lifeless dead.[8]

We have very little definite information about the nature of metempsychosis. One problem is that we have very little on Pythagoras' account of the soul and we do not know if the entire soul or only part of it was supposed to transmigrate. We have nothing at all on the nature of the actual transmigration, of how the soul moved from its previous host body to the next host body. We do not know if every soul underwent transmigration, we do not know the extent of how many living things could participate (animals other than dogs, plants?), and we do not know if there was eventually an escape from the sequence of transmigration, either by death of the soul or escape to some heaven or state that did not involve embodiment.[9]

Shamanism?

It has been suggested that either Pythagoras was a shaman, or that what he did was related to shamanism. A shaman is someone who enters into a state of altered consciousness (perhaps induced by drugs, meditation or repetitive music/dance) and then claims to be able to commune with or perhaps in some manner affect or control the souls of the dead. The social anthropologist Shirokogoroff, who was one of the first to investigate the shaman of the Siberian Tungus people, said that:

In all Tungus languages this term (saman) refers to persons of both sexes who have mastered spirits, who at their will call and introduce these spirits into themselves and use their power over the spirits in their own interests, particularly helping other people, who suffer from the spirits; in such a capacity they may possess a complex of special methods for dealing with the spirits.[10]

The notion of a trance, or some form of ecstatic state, leading to access to a spirit world is the key part of shamanism. There is, though, no reliable evidence that Pythagoras entered trances or ecstatic states and the notion of entering a spirit world is contrary to the principles of metempsychosis. If souls do not enter into some sort of afterlife, but transmigrate to other bodies, what spirit world is there for Pythagoras to enter via some form of ecstatic state? It is perhaps significant that within shamanism proper there is no trace of any view like metempsychosis.

How to live better

Pythagoras was most famous in the ancient world for specifying how to live better. So we can find Isocrates saying that Pythagoras:

More conspicuously than others paid attention to sacrifices and rituals in temples.[11]

We know little of precisely what Pythagoras prescribed here, only that he paid keen attention to these matters. One part of this better way of life was vegetarianism, although it is not clear whether the Pythagoreans were outright vegetarians or only refused to eat certain types of meat. The evidence here is confused, some saying that Pythagoras would not even go near butchers and hunters, others saying that Pythagoras would not eat some parts or types of animals but would eat others. It may well be that the vegetarianism was related to the belief in metempsychosis with the ban on eating certain animals related to which animals were able to partake in metempsychosis. The Pythagoreans were also forbidden from eating beans. The reason for this may be simple and crude – that flatulence is not very helpful if people, either individually or in a group, are meditating and attempting to reach some higher plane of consciousness. Alternatively, it has been suggested that this is related to shamanism as some shamans refuse to eat beans.[12]

Certainly for some Pythagoreans there was a ban on suicide, again perhaps related to the issue of metempsychosis and the best way to enter the next life. Theories of this type often held that what you did in this life determined the nature of your next life and that there was a hierarchy of incarnations. It also seems that some Pythagoreans believed the soul to be in some sense a harmony or attunement. The nature of the soul and its fate after death are an important theme in Plato's Phaedo where some Pythagorean ideas are discussed. The question of how to live for these Pythagoreans was then one of how to bring one's soul into better harmony or attunement. This brings us back to number again as the Pythagoreans are associated with the idea that we can express musical harmony in terms of number.

Mathematics and music

Pythagoras is sometimes credited with the first application of mathematics to music theory. The general idea is straightforward. If we have a stringed instrument, we get a certain note when that string is played 'open'. If we alter the effective length of the string, we can get different notes. The discovery attributed to Pythagoras is that using ratios of simple integers to determine where to stop the string, we can produce harmonious notes. So a ratio of 2:1 will produce an octave, while 4:3 will produce a musical fourth, and 3:2 will produce a musical fifth. We have no direct evidence that Pythagoras discovered this and there is no application or development of this discovery which is attributed to Pythagoras himself. As we will see later, both Philolaus and Archytas, followers of Pythagoras, made considerable use of this insight in developing musical theory. There are many tales about Pythagoras' discovery, but all of these are apocryphal and often physically impossible. Figure 2.1 manages to give five physically impossible ways for this discovery to have been made. Treating these clockwise, starting from the top left:

1. One popular tale had Pythagoras discovering the musical ratios when passing a blacksmith's shop and supposedly noticing that different sized hammers produced different notes. However, the weight of hammer has no direct relationship to the note produced when it hits something. Try it yourself if you like!
2. The size or weight of a bell has no direct relationship to the note it will produce when struck.
3. The amount of water in a glass has no direct relationship to the note produced when the glass is struck.
4. Tensioning strings with different weights looks the most plausible of the methods proposed here. However, changing string tension by using differing weights again does not produce the required relationship to the pitch of the open string (frequency varies in proportion with the square root of the tension). It is only by stopping strings that the required ratios are generated.
5. There is of course a relationship between length of pipe and note produced, but once again it is not the relationship shown here.

Figure 2.1 A late medieval woodcut from Franchino Gafurio's *Theorica musice* (1492) giving five apocryphal representations of how Pythagoras linked music and mathematics.

That someone among the Pythagoreans discovered these ratios – or more likely, in discussion with musicians realized the significance of these ratios – is beyond doubt. As we will see, some Pythagoreans made major contributions to music theory. However, there is no direct evidence that Pythagoras himself had anything to do with this. It is very likely that the discoveries of the later Pythagoreans were attributed to Pythagoras himself in some of the later sources.[13]

Tetraktys

It is likely that the idea of the tetraktys can be traced back to Pythagoras. The tetraktys is the first four integers and their sum is the Pythagorean perfect number, 10 (1 + 2 + 3 + 4). There are records of a Pythagorean oath as:

No, I swear by he who gave to our heads the tetraktys,

The origin and root of immortal nature.[14]

The first four integers were arranged in this manner to form the tetraktys shown in Figure 2.2.

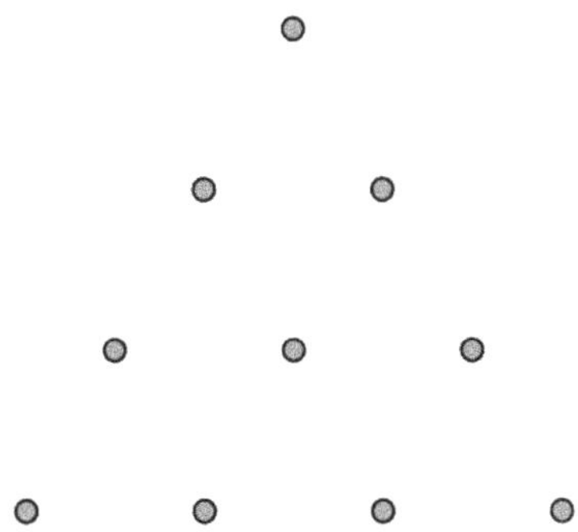

Figure 2.2 The tetraktys links visually the first four integers and the Pythagorean perfect number 10.

The Pythagoreans were quite keen on representing numbers in this way and, as we shall see later, they were also keen on using representations like this in order to understand the relations between numbers. The first four integers and the Pythagorean perfect number 10 feature prominently in Pythagorean thought. Some Pythagorean theories of music derive ratios related to musical notes which use only these first four integers. There is a Pythagorean cosmology where it is supposed that there are 10 (the perfect number) objects orbiting around a central fire.

A world of number?

It has sometimes been said that the Pythagoreans considered the world to be constituted out of numbers. It has never been entirely clear how to visualize this theory, but there is one contrast that may throw some light on this. While the Pythagoreans have been said to have an arithmetical cosmology, Plato has been said to have a geometrical cosmology. That is, while the Pythagoreans considered the world about us to be constituted from numbers, Plato considered it to be constituted from shapes.[15] So for Plato there were two fundamental triangles, which formed either a more complex triangle or a square, which in turn formed three dimensional shapes: tetrahedron, octahedron or icosahedron from the complex triangles or a cube from the squares. These shapes were fire, air, water, and earth respectively.

Philosopher of Science Karl Popper has commented that one of Plato's main contributions is that:

> Ever since Plato and Euclid, but not before, geometry (rather than arithmetic) appears as the fundamental instrument of all physical explanations and descriptions, in the theory of matter as well as cosmology.[16]

So one might say that since Plato we have thought in terms of geometrical shapes for the fundamental particles that make up our world. We have thought in terms of spherical atoms and when atoms were discovered to be comprised of smaller particles, we have thought in terms of spherical electrons, protons, and neutrons. It might also be said that the twentieth century has re-instated a more Pythagorean picture, as with the advent of quantum mechanics we now think of electrons in terms of wave or probability functions rather than in terms of shapes.

Modern scholarship on the Pythagoreans has moved on slightly though.[17] It is remarkably difficult actually to find any Pythagorean who explicitly advocated the idea that the world about us is indeed constituted from numbers. There is little evidence for this prior to Aristotle, who tells us that:

> Contemporaneously with these philosophers and before them, the so-called Pythagoreans, who were the first to take up mathematics, not only advanced this study, but also having been brought up in it they thought its principles were the principles of all things. Since of these principles numbers are by nature the first, and in numbers they seemed to see many resemblances to the things that exist and come into being.[18]

Aristotle also says that:

> The Pythagoreans believed in one kind of number, the mathematical. They hold that it is not separate, but sensible substances are constituted out of it. They construct the whole heaven out of numbers, not abstract units, but units which have size. However, on the subject of how the first extended one is constructed, it is likely that they are in difficulty.[19]

There are many ways though in which one might think that numbers are important in giving an account of the world without actually believing that the world is literally constituted from numbers.

Pythagoras' powers?

There are many tales of strange powers and deeds associated with Pythagoras. Some mentioned by Aristotle are that he was seen in two different places at exactly the same time, that one of his thighs was golden, and that when crossing the River Kosas the river spoke to him and many people heard this.[20] Other tales have Pythagoras prophesying the coming of a white, female bear, killing a dangerous snake by biting it, and to have prophesied to his followers approaching political strife.[21] We have no direct evidence of Pythagoras making any of these claims himself nor do we know his attitude to any of these claims.

What we make of these tales is open to debate. One view is that it is not surprising that these sorts of tales were attributed to a secretive, charismatic religious leader in antiquity and we need not take them too seriously. A second view is that these tales are in some way symbolic and have considerable significance in terms of magic, ritual, and access to the realm of the dead, and they fit into a broader pattern of such stories.[22]

Acousmatikoi and *mathematikoi*

There were two groups of immediate followers to Pythagoras', the *acusmatikoi* and the *mathematikoi*, the listeners and the learners. The classic statement of the division between the followers of Pythagoras is given by classics scholar F.M. Cornford, who says:

> Tradition points to a split between the Acousmatics, who may, perhaps, be regarded as the 'old believers' who clung to the religious doctrine, and the Mathematici, an intellectualist or modernist wing, who, as I believe, developed the number doctrine on rational, scientific lines, and dropped the mysticism.[23]

However, it is doubtful that such a bipolar split can be justified given more modern historiographies and it is more likely that there was a much wider spectrum of views, including these two wings but also those who brought together the religious, magical, and scientific aspects. Pythagoreanism was more of a broad church where some may have felt happy with a mix of what Cornford categorizes here as religious and scientific views. Cornford wrote this in the 1920s when ideas of an inherent conflict between religion and science, and indeed magic and science, were much more prevalent than they are today. So too ideas of a linear progression for humanity from magic to religion to science were more prevalent.

Hippasus and √2

One of the more colourful stories relating to the early Pythagoreans is that of Hippasus and the irrationality of the square root of 2. We know very little of Hippasus, other than that he was associated with the Pythagoreans and lived in the fifth century BCE. According to some tales, he may have discovered the irrationality of $\sqrt{2}$. A rational number is one that can be expressed as the ratio of two integers. An irrational number is one that cannot. It is said that having discovered the irrationality of $\sqrt{2}$, Hippasus also made this generally known and was then drowned at sea. It is believed that irrational numbers were discovered around this time, though in fact we have very little evidence on this. The irrationality of several square roots was certainly known to Plato, as is clear in his *Theaetetus*. How and why Hippasus was drowned at sea, if that indeed was his fate, is the subject of several stories. There is a good deal of variety and contradiction among these stories and it is difficult tell which, if any one of them, is true. The basic idea is that the discovery of the irrationality of $\sqrt{2}$ was in some way embarrassing and that Hippasus' death by drowning was in some way a punishment for either discovering or divulging the irrationality of $\sqrt{2}$. The embarrassment to the Pythagoreans is supposed to be that their belief in a cosmos comprised of numbers, where all relations should be expressible as the ratio of two integers, was compromised by the discovery of the irrationality of $\sqrt{2}$. How much of a problem that was for the Pythagoreans will depend on how much they were committed to the idea of a cosmos comprised of numbers. So either for the discovery of the irrationality of $\sqrt{2}$, or for divulging this knowledge contrary to Pythagorean principles of secrecy, Hippasus was in some way put to death by drowning by the Pythagoreans. Some tales have Pythagoras ordering this, some have Hippasus discovering the irrationality of $\sqrt{2}$ while on a sea voyage and being thrown overboard by his Pythagorean fellow travellers. Other versions of the tale have the gods as implicit in Hippasus' death by drowning, or the revealed secret being how to construct a dodecahedron inside a circle. Yet more versions have Hippasus merely being expelled from the Pythagorean brotherhood.

There are issues with this story apart from the multiple and conflicting versions of it. First, why should the Pythagoreans find the discovery of the irrationality of $\sqrt{2}$ so embarrassing? Only if they were committed to the idea of the world being comprised of integers and all relations being expressible in terms of ratios of those integers would this be problematic. However, as we have seen, there is very little evidence to tie the Pythagoreans to such a belief.

Pythagorean numerology

The Pythagoreans are supposed to have had an interest in numerology. There is a need here to be careful about what sort of numerology is ascribed to which

Pythagoreans, as there are many types of numerology and as we have seen Pythagoreanism was quite a diverse phenomenon encompassing a good many different ideas and attitudes. The first thing to say is that while there is something in the modern world called 'Pythagorean Numerology', as far as we know this sort of numerology was not practised by the Pythagoreans. The basic idea of modern Pythagorean numerology is that we can tell something about someone's character or fate by substituting numbers for the letters of their name (a = 1, b = 2, etc.) and then manipulating those numbers, along with the numbers of their birthday, to reach a single figured integer. So for simplicity, let us take someone called Aaron Abbs, born 01/02/2000.

$$\text{Aaron} = 1 + 1 + 18 + 15 + 14 = 39$$
$$3 + 9 = 12$$
$$1 + 2 = 3.$$
$$\text{Abbs} = 1 + 2 + 2 + 19 = 24$$
$$2 + 4 = 6.$$
$$01/02/2000 = 1 + 1 + 2 = 4.$$

Aaron Abbs then has key numbers of, 3, 6, and 4. If 3 = motivated, 6 = strong, and 4 = artistic, Aaron Abbs is motivated, strong, and artistic. The process of manipulating the numbers here is entirely arbitrary and can be made more complex and the interpretation of the key numbers made more complex to give the numerologist an air of expertise or mystery, but the basic principles remain the same. However, much as modern numerologists would like to give their practice ancient authority or mystique by linking it to early Pythagoreans, they were simply not numerologists of this sort.[24]

This is not to say that the Pythagoreans were uninterested in what we might term numerology. This is probably better put by saying that the Pythagoreans were interested in all aspects of numbers and their properties. They did not really have a distinction between a purely mathematical property of a number and numerology, which is not surprising in their historical context. Indeed, it is not as easy as it might seem, even today, to give a watertight definition of what is mathematical and what is numerological. Modern philosopher David Stove has commented that:

> No one actually knows, even, what is wrong with numerology. Philosophers, of course, use numerology as a stock example of thought gone hopelessly wrong, and they are right to do so; still, they cannot tell you what it is that is wrong with it. If you ask a philosopher

this, the best he will be able to come up with is a bit of Positivism about unverifiability, or a bit of Popperism about unfalsifiability.[25]

While it is easy to rule out the 'Pythagorean' numerology we have seen above, the fact is that numbers do have some interesting properties and it is not so easy to draw the line between the mathematical and the numerological – and it certainly would not have been easy for the Pythagoreans.

If we want a slightly surprising property of numbers, which was certainly known to the Pythagoreans, there is the fact that, loosely put, the sums of odd numbers give the square numbers. Put more precisely, the sum of the first n odd numbers = n^2 So:

$$1 = 1 = 1^2$$
$$1 + 3 = 4 = 2^2$$
$$1 + 3 + 5 = 9 = 3^2$$
$$1 + 3 + 5 + 7 = 16 = 4^2$$
$$1 + 3 + 5 + 7 + 9 = 25 = 5^2$$

This is shown pictorially in Figure 2.3.

Some numbers are 'perfect' numbers where 'perfect' is a technical term and the number is equal to the sum of its proper divisors. 6 is divisible by 1, 2, and 3.

6 = 1 + 2 + 3, making 6 a perfect number.

There is a sequence of perfect numbers, the next being 28:

$$28 = 1 + 2 + 4 + 7 + 14.$$

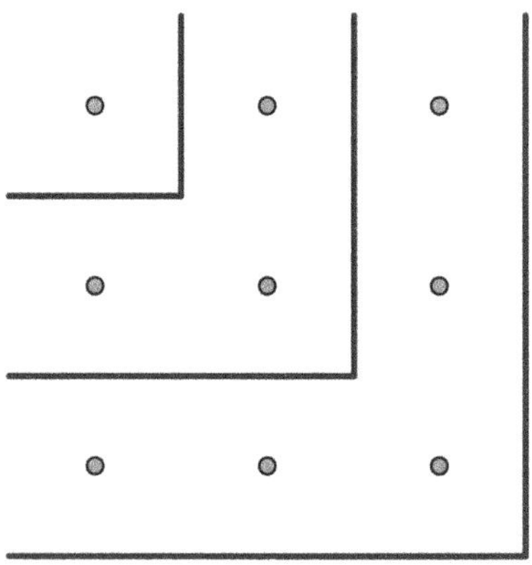

Figure 2.3 Placing counters in the shape of a square illustrates that the sum of the first n odd numbers is n^2.

496, 8128 are the next two and there are currently 48 known perfect numbers. The Pythagoreans were aware of the idea of perfect numbers and had the terms 'under-perfect' and 'over-perfect' for numbers whose factors added up to less or more than the number respectively.

That some Pythagoreans were interested in what we would call numerology is undeniable. They did attribute non-mathematical properties to numbers. So 2 and 3 were associated with male and female, while 5 was associated with marriage and 10 was seen as a divine or special number. There are some important points to make here, though. Numbers do have properties and it would not have been easy for the Pythagoreans, in the context of what was known in ancient Greece, to distinguish what we would consider mathematical and numerological properties. There was a wide spectrum of interest in number among the Pythagoreans. Some would have been what we would see as mathematicians, some what we would see as numerologists and some would have mixed aspects of these two extremes together. Many ancient societies had forms of numerology where there were numerically good days to do things and bad days to do things (superstition about Friday the 13th is a hangover of this sort of thinking).[26] Did the Pythagoreans go beyond this sort of thinking? I would suggest they did, not only in their purely mathematical thinking, but in some of their forms of numerology as well.

What we find in the Pythagoreans, but not in other early cultures, is the attempt to apply what are thought to be certain privileged numbers, the tetraktys of 1, 2, 3, and 4 or the perfect number 10 generated from the tetraktys to draw conclusions about the nature and structure of the heavens or the world. So there are 10 bodies in the heavens because that is the perfect number and their motion is related to the tetraktys because that is related to the celestial music.

Philolaus on music

Philolaus of Croton lived from *c.* 470 to *c.* 385 BCE. Philolaus and Archytas were the most significant contributors to the Pythagorean tradition we know of in the presocratic period.

Philolaus wrote one book, *On Nature*, which if Pythagoras wrote nothing is probably the first book of the Pythagorean tradition, of which a few fragments survive. He worked on astronomy, cosmology, and music theory.

Philolaus did important work on the mathematical theory of music and harmony. He was aware of the basic Pythagorean ideas of a 2:1 ratio for an octave,

4:3 for a musical fourth, and 3:2 for a musical fifth. Philolaus introduced the ratios of 9:8 for a whole note and 256/243 for a semitone. The way this works is that if we take our root note as 1, then the next whole note is 9/8 (1 x 9/8). The following note is then 9/8 x 9/8 = 81/64. So using the key of C major, the Philolaus values for the notes can be generated as in Table 2.1.

Table 2.1 ***Philolaus' scale. The first row gives modern note names, the second row is the ratio between notes, the third row is the note expressed as a ratio relative to the root note.***

C	D	E	F	G	A	B	C
	9/8	9/8	256/243	9/8	9/8	9/8	256/243
1	9/8	81/64	4/3	3/2	27/16	243/128	2

While the 256/243 ratio looks a little complex, each number here is based on a tetraktys number. So 9/8 is 3 × 3/2 × 2 × 2, 81/64 is 9/8 × 9/8, 27/16 is 3 × 3 × 3/2 × 2 × 2 × 2, and even the arbitrary looking numbers as 128, 256, and 243 are powers of 2 and 3. 128 = 2^7, 256 = 2^8, and 243 = 3^5 (= 3 x 81, when 81 = 3 × 27 and 27 = 3^3).

In modern musical theory, we have something called 12-tone equal temperament (12ET), where for an octave there are 12 equally sized semitones. The ratio between all neighbouring semitones in 12ET is $\sqrt[12]{2}$ (the 12th root of two). Pythagorean scales (also known in modern terminology as 'just intonation') do not have this property. Both the ratios between notes and the position of the notes within the octave can be expressed in terms of 'cents'. Ratios are said to be '100 cents' when they match the 12ET ratio, where 1200 cents make up one octave. Positions are said to be 100 cents when they match the 12ET positions. The differences between 12ET and Philolaus are shown in Table 2.2.

Table 2.2 ***Philolaus against the modern scale. The first row gives modern note names, the second row is equal tempered notes expressed in cents, and the third row is Philolaus notes expressed in cents.***

C		D		E	F		G		A		B	C
0	100	200	300	400	500	600	700	800	900	1000	1100	1200
		203.91		407.82	498.04		701.96		905.87		1109.78	1200

There are advantages and disadvantages to modern 12ET. The advantages are that keyboards tuned to 12ET, and instruments with the frets placed according to 12ET (standard modern guitars), can be played in any key equally well without retuning. This allows key changes within one piece of music much more easily and also facilitates ensemble playing. Chords also sound better in 12ET, especially the more complex chords used for jazz.

The disadvantage of 12ET is that the harmonies produced do not sound quite as pure as those of the Pythagorean scale. So Pythagorean harmonies are still sometimes used where tuning the instrument is not an issue, as for example with the human voice, where barber shop quartets can make use of Pythagorean harmonies.

Cosmology

Philolaus gave us a specific model of the heavens shown in Figure 2.4. In many ways this is a remarkable model of the heavens for antiquity. Moving outwards

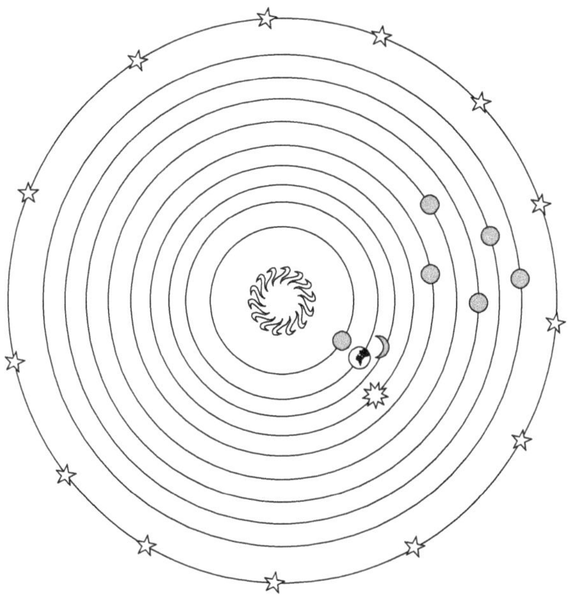

Figure 2.4 Philolaus' model of the heavens. Moving outwards from the middle, the celestial bodies are a central fire, a counter-earth, the earth, the moon, the sun, the five naked-eye planets (Mercury, Venus, Mars, Jupiter, Saturn), and the stars. The counter-earth prevents the central fire being visible from earth.

from the middle, the celestial bodies are a central fire, a counter-earth, the earth, the moon, the sun, the five naked eye planets (Mercury, Venus, Mars, Jupiter, Saturn), and the stars. One reason that this is remarkable is that it is one of the very few ancient models of the heavens which has the earth in motion, rather than immobile and at the centre of the cosmos. It is also remarkable that it is not the sun that is at the centre of the cosmos in place of the earth, but a central fire. No reason has been passed down to us as to why the earth was placed in motion. We cannot see the central fire as the counter-earth is in a synchronous orbit with the earth and always blocks our view of the central fire. How well this model could account for the phenomena is still open to debate. Aristotle has a rather critical view of this cosmology:

The Italian philosophers known as Pythagoreans take the contrary view. At the centre, they say, is fire, and the earth is one of the stars, creating night and day by its circular motion about the centre. They further construct another earth in opposition to ours to which they give the name counter-earth. In all this they are not seeking for theories and causes to account for observed facts, but rather forcing their observations and trying to accommodate them to certain theories and opinions of their own.[27]

This leads us into another famous Pythagorean idea, the 'harmony of the spheres'.

Music of the spheres

Aristotle gives us several passages on the Pythagoreans and music in the heavens. He tells us in the *Metaphysics*:

Since, again, they saw that the modifications and the ratios of the musical scales were expressible in numbers;-since, then, all other things seemed in their whole nature to be modelled on numbers, and numbers seemed to be the first things in the whole of nature, they supposed the elements of numbers to be the elements of all things, and the whole heaven to be a musical scale and a number. And all the properties of numbers and scales which they could show to agree with the attributes and parts and the whole arrangement of the heavens, they collected and fitted into their scheme; and if there was a gap anywhere, they readily made additions so as to make their whole theory coherent. E.g. as the number 10 is thought to be perfect and to comprise the whole nature of numbers, they say that the bodies which move through the heavens are ten, but as the visible bodies are only nine, to meet this they invent a tenth — the 'counter-earth'.[28]

We do not know how the Pythagoreans applied their music theory to the heavens, only that they did. When asked the question of why we cannot hear this celestial music, they replied that, just as workers in a smithy do not hear the beating of the hammers, we do not hear the celestial music. It is part of the background of the cosmos and we are simply too accustomed to it to be able to hear it.

Later in antiquity, Plato also adopted the idea of a celestial music in his *Republic*. His *Timaeus* is also interesting in that he uses the ratios of Philolaus' music theory to give the spacing between the celestial bodies. Much later than this, in the late sixteenth and early seventeenth centuries, the great astronomer Johannes Kepler took up this idea, this time with the sun in the centre and the earth and planets orbiting it.

Archytas

Archytas of Tarentum (*c.* 428 BCE–*c.* 347 BCE), another early Pythagorean, was important for his work in mathematics, cosmology, and music theory. Plato clearly treats Archytas as a Pythagorean when he says:

> As our eyes are suited to astronomy, so our ears are suited to harmony, for these are brother disciplines, as the Pythagoreans say and we, Glaucon, agree.[29]

There is a similar grouping of the sciences to Archytas. Archytas does not seem to have been interested in the idea of metempsychosis, nor is there any sense of mysticism or numerology. Archytas was extremely concrete in everything he said and clearly regarded the art of calculation, or logistic, as the key science. So Archytas says:

> It seems to me that those concerned with the science make distinctions well and it is by no means surprising that they understand individual entities as they are. Having made good distinctions concerning wholes they are also able to see well how things are according to their parts. Concerning geometry, arithmetic, and spherics he gave clear distinctions and not least concerning music. These sciences seem to be akin.[30]

Elsewhere he states;

> It seems that logistic is far better than the other crafts in respect of wisdom and deals with its topics more concretely than geometry. In those ways in which geometry is lacking logistic utilises demonstration.[31]

Archytas on cosmology

There is an interesting thought experiment in cosmology that is attributed to Archytas which was much debated and was very influential in antiquity.[32] If someone were to stand close to the edge of a finite cosmos and attempted to thrust a staff beyond the edge of the cosmos, what would happen? If they succeed, then this cannot be the limit of space, and so there must be a new limit further on. This thought experiment though is infinitely replicable. Wherever a new edge is supposed we can imagine someone standing close to it and thrusting a staff beyond it. So space must be infinite. One reply is physical and practical, that it is impossible to stand close enough to the edge of the cosmos in this manner and so the idea of thrusting a staff beyond the edge of the cosmos is impossible too. A more subtle reply is that outside the cosmos neither time nor space exist and it is impossible to thrust the staff where there is no time or space. Our intuition that we thrust the staff beyond the edge of the cosmos is incorrect and so space is finite after all.

Archytas and mathematics

Archytas worked on one of the notorious problems for ancient mathematics, the Delian problem, of doubling the volume of a cube.[33] While initially this looks simple, in fact it is very tricky, especially within the confines of the mathematical techniques then known to the ancients. Archytas developed the work of Hippocrates of Chios. If we suppose that L is the length of a side of the original cube, one can then generate a series of ratios such that L:a:: a:b:: b:2L.[34] One can derive the relation $L:2L = L^3:a^3$. As $L^3:a^3$ is in the ratio of 1:2, a^3 is twice L^3, and the cube can be generated with sides of length a. Archytas' solution is too complex to give here. It involved constructing four similar triangles in the proportions suggested by Hippocrates, employing an imaginary rotation of triangles and then calculating of their points of intersection. Archytas' solution is one of the most remarkable pieces of technical mathematics, visualization and mathematical ingenuity in antiquity.

Archytas also demonstrated a very important property of what are known as superparticular ratios. These ratios are of the type where n + 1:n. These were important for Pythagorean musical theory, which used ratios such as 3:2, 4:3, and 9:8. If x bears the same proportion to y as y does to z, then y is the mean proportional of x and z (if x:y:: y:z). A double octave (4:1) can be split into two octaves with a mean proportional as 4:2 is the same proportion as 2:1. Archytas demonstrated that there is no mean proportional for numbers in superparticular ratios. This means that critical musical ratios, such as 3:2, 4:3, and 9:8 (which all have the form n + 1:n) have no mean proportional and cannot be split in to two equal parts.

Archytas and music theory

Archytas produced a variation on Philolaus' musical scale, using 9:8, 8:7, and 28:27 to generate the notes up to the fourth (9/8 × 8/7 × 28/27 = 4/3). Archytas worked on two other types of scale, in modern terminology the chromatic and the enharmonic. A chromatic scale includes all twelve semitones. The key ratios for Archytas' chromatic scale are 32:27, 243:224 and 28:27 (32/27 × 243/224 × 28/27 = 4/3). In the chromatic scale, $A^{\#} = B^{b}$. In an enharmonic scale this is not so, and what we would call $A^{\#}$ differs from B^{b}. The key ratios for Archytas' enharmonic scale are 5:4, 36:35, 28:27 (5/4 × 36/35 × 28/27 = 4/3).

In contrast to Philolaus, who seemed to be producing an ideal scale, Archytas is now generally agreed to have been describing the scales in use during his time.[35] With Philolaus, certain numbers are privileged by either being part of, or directly derivable from, the tetraktys. This is not the case for several of the ratios which Archytas uses, such as 8:7, 28:27, 32:27, 243:224, 5:4, and 36:35. One might argue that to some extent Philolaus had a numerological approach to music theory, while Archytas did not. Archytas also had a physical theory of pitch.[36] The pitch of a sound in this theory is related to how quickly it travels, a sound travelling more quickly being of higher pitch. In fact the speed of sound is a constant for a given medium and it is frequency that is critical to pitch, how rapidly a string vibrates determining the frequency rather than the speed of the sound.

Plato

Plato (428/427–348/347 BCE) is important both as a philosopher and as someone who was interested in and promoted the study of mathematics. In the philosophy of mathematics the view of 'Realism', or 'Platonic Realism' – the idea that numbers have an independent existence apart from the things they count – dates from Plato. Plato used mathematics as a model for other types of knowledge. It is said that the words 'Let no one ignorant of geometry enter here' were inscribed over the entrance to Plato's academy, the research school which he founded. Plato also gave mathematics and geometry a critical role in the education of the guardians of Plato's ideal state.

Whether Plato was himself a Pythagorean, or whether some of his dialogues should be considered as Pythagorean, has been a matter of some debate. Plato was clearly knowledgeable about both Pythagoras and Pythagorean ideas and it is clear he was to some extent influenced by Pythagorean ideas. It is now widely accepted that it would be misleading to consider Plato an outright Pythagorean, and that while there is an influence, it would also be incorrect to consider any of his dialogues to be simply Pythagorean, either in derivation or in content. In the twentieth century the classicist A.E. Taylor held that Plato's *Timaeus* was derived from Pythagorean sources, but this view is now largely discarded.[37] It is very important in considering the relation between Plato and the Pythagoreans not to use blanket terms like 'number mysticism'. The dominance of positivist and empiricist ideas in the twentieth century, with its rejection of anything which was not either logically true or empirically verifiable, tended to blur the differences between Pythagorean and Platonic approaches to number.

Let us begin with some important differences between Plato and the Pythagoreans. Plato did not accept the notion of the soul as a harmony.[38] In terms of cosmology, the cosmos of Plato's *Timaeus* is finite and bounded where Archytas argued for an infinite, unbounded cosmos.[39] The *Timaeus* also has a very different account of the number, nature, motions, and organization of the celestial bodies from that of Philolaus.[40] There are two other important structural differences between Plato and the Pythagoreans in approaching cosmology, the first of which we have seen something of already.

One can broadly categorize the Pythagorean approach as arithmetical.[41] That is they were interested in the relation of numbers to the world. Plato had a much more geometrical conception of the cosmos. There are the 1, 1, $\sqrt{2}$, and 1, $\sqrt{3}$, 2 triangles from which the cubes of earth and the tetrahedra, octahedra, and ikosahedra of fire, air, and water are formed (see Figure 2.5). It is these shapes that form the basis of Plato's cosmos, not numbers themselves.

A second issue is the generation of the musical scale. Philolaus' scale uses the tetraktys of 1, 2, 3, and 4 to generate its ratios, the justification being that these are part of the tetraktys and that $1 + 2 + 3 + 4 = 10$, the Pythagorean perfect number. While the Pythagoreans have ten heavenly bodies, Plato simply accepts there are seven heavenly bodies (moon, sun, five naked-eye planets) and has seven terms as basic to his musical scale (1, 2, 3, 4, 8, 9, 27),[42] which are the

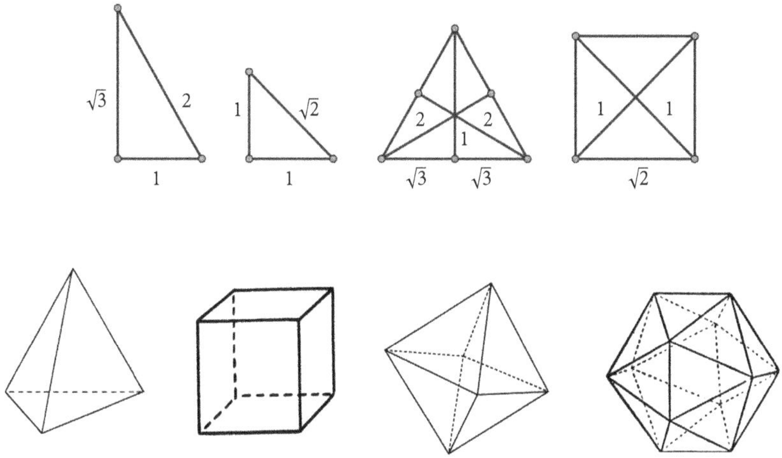

Figure 2.5 (a) Plato's two fundamental triangles, which can be used to construct the equilateral triangle, square, and thence (b) the tetrahedron, cube, octahedron, and icosahedron.

relative lengths of the soul stuff that the demiurge, Plato's geometer/craftsman god, uses to fashion the orbits for these bodies.[43] Plato then generates a tone and semitone scale from these terms.[44] Again, the derivation is geometrical (dividing the soul stuff into circles) rather than purely arithmetical as with the Pythagoreans. So while the Pythagoreans have a numerological derivation of cosmology and of music, Plato has a cosmological derivation of music.

A very important part of Plato's thinking in his *Timaeus* is the idea of the demiurge who organizes the best possible cosmos from a pre-existing chaos. In order to make the cosmos good and comprehensible to humans, this god imposes number and geometrical figure onto the chaos, which is why, for example, the fundamental particles for Plato have geometrical form. One important aspect of this cosmology is that the demiurge needs criteria for all that he does and he finds those criteria in mathematics and geometry. While we might find it strange that one form of triangle should be better than another, Plato did not:

This we hypothesise as the principle of fire and of the other bodies . . . but the principles of these which are higher are known only to God and whoever is friendly to him. It is necessary to give an account of the nature of the four best bodies, different to each other, with some able to be produced out of the others by dissolution . . . We must be eager then to bring together the best four types of body, and to state that we have adequately grasped the nature of these bodies. Of the two triangles the isosceles has one nature, the scalene an unlimited number. Of this unlimited number we must select the best, if we intend to begin in the proper manner. If someone has singled out anything better for the construction of these bodies, his victory will be that of a friend rather than an enemy. We shall pass over the many and postulate the best triangles.[45]

If we ask how the heavens should be arranged, what the ratios should be between the orbits of the planets, then Plato's answer is that the demiurge does this using a musical scale. That produces the best arrangement and one that is mathematically comprehensible to humans. This line of thought was both important and long lasting. One can find Johannes Kepler following it in the late fifteenth and early sixteenth centuries. For Kepler, the question is why the orbits of the planets have certain ratios, and why there are this many planets. A second issue is why the elliptical orbits of the planets have specific eccentricities and the planets specific speeds. His answer to the first question is geometrical. The planets have the same ratios as constructions of the five known Platonic solids, and there are this number of planets because there are this number of Platonic solids. His answer to the second question is musical. The planets have these eccentricities and speeds because those values conform to a celestial music. For an explicitly Christian aspect to Kepler's astronomy, see Owen Gingerich's Chapter 4 in this volume.

The early Pythagoreans and cosmic number

The early Pythagoreans are different from Plato and Kepler in that they did not believe in a creator god organizing the cosmos, as far as we can tell from the evidence. So there is no issue for them concerning the criteria by which a creator god did this, even if there was for figures like Plato and Kepler. However, they were interested in the application of number and music theory to the cosmos and how that might make the cosmos good, and make the cosmos comprehensible to humans. Philolaus says that:

Nature in the cosmos was fitted together out of unlimited and limited things, both the whole cosmos and the things in it.[46]

Who or what fitted the cosmos together from unlimited and limited things is not an issue here.[47] The key is that there is a difference between a pre-cosmic state and the cosmos for the Pythagoreans and it relates to harmonization. Philolaus elsewhere states that:

The first thing to be fitted together, the one, is in the middle of the sphere and is called the hearth.[48]

In order for a cosmos to be a proper cosmos, it must have a mathematical structure. That may not be quite so strong or explicit as in Plato here, but nevertheless there is still an important sense of it. I have translated the Greek somewhat conservatively here as 'to fit together', but the Greek word also has a musical sense of to bring into tune, as in to tune an instrument, or even to compose music. It also has a moral sense above simply (as, say with a carpenter) an appropriate or suitable fitting together. Another statement by Philolaus is also important here:

> Concerning nature and harmony they hold this: The being of objects, eternal being and nature itself are susceptible to divine but not human knowledge, although it is not possible for any of the things which are and are known by us to have been generated if the things out of which the cosmos was put together, both the limited and the unlimited things, had not existed beforehand. However, as these origins did exist and were neither alike nor of the same kind, it would not have been possible for them to have been ordered if harmony had not, in some way, been applied to them. On the one hand like things and things of the same kind were not bonded by harmony, while unlike things, things not of the same kind and things not corresponding in order needed to be closed up tightly in harmony if they were to remain held fast in order.[49]

And further:

> All things which are known have number. Without this, it is not possible for anything at all to be understood or known.[50]

So harmony is critical to the structure of the cosmos and number is critical for the possibility of human knowledge of the cosmos.

Let us return to some aspects of numerology that we discussed earlier. The Pythagoreans, and indeed Plato, have been said to employ numerology and 'number mysticism'. In a sense that is true, as both employ numbers in their account of the cosmos in a way that would not be recognized as mathematical physics today. They use privileged numbers and attempt to say how these led to, or constitute a good arrangement of, the world. However, this was not a simple or primitive numerology. They had important philosophical reasons for their application of number. They needed to explain how the cosmos was good, that is in good order and aesthetically good, a standard assumption among the ancients, and how humans could have knowledge of the cosmos. From Plato onwards, there is an assumption of a god organizing the cosmos. For each of the actions of that god there has to be a reason and some of those reasons are supposed to be mathematical or geometrical.[51] This is a far cry from the modern view of an accidental universe where we fit mathematics to what we observe as best we can. It is important to understand that the ancients asked a different question about cosmology, which was how has this all come about for the best, given that the cosmos is so congenial to human beings and appears comprehensible to humans as well?

Plato and a Pythagorean code?

One interesting recent development on the relation between Plato and Pythagoras has been the work of historian of science Jay Kennedy.[52] His claim is that there is a musical structure to each of Plato's works, based on Pythagorean principles, and that this structure effectively contains a code, revealing information about Pythagoras and about Plato's own views. Kennedy claims that Plato organized his work stichometrically, that is he was aware of the number lines in each of his works and that Plato divided each of his works into 12 parts, corresponding to the 12 notes in a musical scale. The claim is then that Plato had means of indicating the transition from one 12th to another, by making a reference to say, divine justice, or a speech by one of his characters may begin at a 12th part of a work. There is also supposed to be a harmonic organization to Plato's works based on a 12-note division of the octave. It is claimed that Plato writes predominantly of ideas he supports at harmonious parts of the scale and predominantly of ideas he does not approve of at dissonant parts. Kennedy claims that Plato's works are 'fundamentally Pythagorean' and that we can find encoded information about Pythagoras in them. Kennedy supports these theses with statistical analysis which at first sight certainly appears impressive in its breadth and claimed results.

If these claims are verified, they will radically change our understanding of both Pythagoras, Plato, and the way that Plato used mathematics. However, this is a big 'if'. There has been considerable debate concerning the methods and assumptions that Kennedy has used. It is not clear that the musical structure Kennedy finds in Plato is an appropriate one for that period. Effectively, Kennedy finds a 12ET structure where one would expect a Pythagorean structure. On the statistical side, there is an important difference between claims for accuracy, which Kennedy makes, and claims for statistical significance. So if I make a similar claim about Shakespeare, that he divided his works into 12 parts and marked the transition from one part to another with the word 'and', this would no doubt be accurate in the sense that 'and' will occur in these places, but not statistically significant. The real test for Kennedy's work would be running proper tests for statistical significance and it is by no means clear that Kennedy's claims would survive such tests.

Conclusion

The idea that Pythagoras himself was an important mathematician, that he discovered or proved Pythagoras' theorem, that he discovered the mathematical ratios underpinning music, that he thought in terms of a world constituted from number or was the originator of the idea of music of the heavenly spheres, has now been rejected. In its place we have a Pythagoras who was primarily interested in the fate of the soul after death, an expert on religious ritual and founder of a religious sect, but who may have been interested in and made some contribution to mathematics as well. Recent changes in how we understand the relation of science and religion and the relation of science and magic can accommodate this new, more interesting picture, one that is more firmly based in the reliable evidence we have about Pythagoras.

The early Pythagoreans, as I have stressed, had a wide diversity of views. Philolaus and Archytas seem much less interested in the fate of the soul and much more interested in mathematical music theory. Philolaus seems to have based his theory firmly in the numbers of the tetraktys, while Archytas seems to have been more interested in what contemporary musicians actually played and how that might be described mathematically. The two broad groups of followers of Pythagoras, the *akousmatikoi* and the *mathematikoi* – the listeners and the learners – also seem to have taken quite different approaches to the teachings of Pythagoras and to the study of mathematics.

Plato is very interesting, both as a figure in the history and philosophy of mathematics in himself and as someone who was influenced by the early Pythagorean tradition. Plato also illuminates some aspects of the Pythagorean use of number in thinking about the structure of the cosmos and how humans can gain knowledge of the cosmos.

It is very important when we look at the early Pythagoreans to place them in their philosophical and scientific context. If we take what they were doing as answers to modern questions, then they come across as interested in an odd number mysticism. If we understand them as looking at ancient questions on the nature of the cosmos, without the benefit of an understanding of modern scientific methods, then they have some interesting and diverse things to say about the application of mathematics to understanding how the cosmos is good, why a god may have organized the cosmos in this way, and how we humans can have some knowledge of that cosmos.

Notes and references

1. The classic discussion here is W. Burkert, *Lore and Science in Ancient Pythagoreanism*, Harvard University Press, 1972, but see also L. Zhmud, *Pythagoras and the Early Pythagoreans*, Oxford University Press, 2012, for the most recent contribution.
2. See C. Huffman, The Pythagorean tradition, in A.A. Long (ed.), *The Cambridge Companion to Early Greek Philosophy*, Cambridge University Press, 1999, 66–87 and also Burkert, *Lore and Science* and Zhmud, *Pythagoras and the Early Pythagoreans.*
3. Burkert, *Lore and Science.*
4. D.W. Graham, *The Texts of Early Greek Philosophy*, Cambridge University Press, 2010 is an excellent recent collection with English translation. On the reports of Plato and Aristotle, and other sources, the classic work is Burkert, *Lore and Science.*
5. Plato, *Republic*, 531a, 600a, *Sophist*, 242c, Aristotle, *Metaphysics* I/v, *De Caelo* II/xiii.
6. Plato, *Republic*, 600b.
7. Diogenes Laertius VIII, 36, Xenophanes Fragment 7. The fragments noted here and elsewhere in this chapter can be found in D.W. Graham, *The Texts of Early Greek Philosophy*, Cambridge University Press, 2010, which is an excellent recent collection with English translation, while G.S. Kirk, J.E. Raven, and M. Schofield, *The Presocratic Philosophers*, Cambridge University Press, 2nd edn, 1983 is the most readily available collection, again with an English translation.
8. Homer, *Odyssey*, XI 489.
9. Cf. Huffman, The Pythagorean tradition, pp. 70–1.
10. S.M. Shirokogoroff, *Psychomental Complex of the Tungus Trubner*, K. Paul, Trench, 1935, p. 269.
11. *Busiris*, 28.
12. See P. Kingsley, 1995, *Ancient Philosophy, Mystery and Magic*, Clarendon Press, 1995.
13. See here Huffman, The Pythagorean tradition, pp. 66–87; Zhmud, *Pythagoras and the Early Pythagoreans.*
14. Pseudo-Plutarch, I, 3, 8.
15. See A.D. Gregory, *Plato's Philosophy of Science*, Duckworth, 2000, pp. 238 ff.
16. K.R. Popper, The nature of philosophical problems and their roots in science, *The British Journal for the Philosophy of Science*, **3**(10), 1952, 124–56, pp. 147–8.

17. See Huffman, The Pythagorean tradition, pp. 66–87, Zhmud, *Pythagoras and the Early Pythagoreans*.

18. Aristotle, *Metaphysics*, I/8, trans. Ross.

19. Aristotle, *Metaphysics*, XIII/6, 1080b16–22.

20. Aristotle Fragment, 191.

21. Aristotle Fragment, 191.

22. P. Kingsley, *Ancient Philosophy*.

23. F.M. Cornford, Mysticism and science in the Pythagorean tradition, *The Classical Quarterly*, **16**(3/4), Jul.–Oct., 1922, 137–50, p. 139.

24. There is no evidence that the Pythagoreans ever used this sort of numerology. See Burkert, *Lore and Science*; A.D. Gregory, *The Presocratics and the Supernatural*, Bloomsbury, 2013.

25. D. Stove, *The Plato Cult and Other Philosophical Follies*, Blackwell, 1991, chapter 7.

26. Concern about Friday 13th seems to be a largely twentieth-century phenomenon and due to an amalgamation of 13 as an unlucky number and Friday as an unlucky day.

27. Aristotle, *On the Heavens*, II/13, trans. Stocks.

28. Aristotle, *Metaphysics*, I/5, trans. Ross, cf. Aristotle, *On the Heavens* II/13 and Aristotle, *On the Heavens*, III/1, trans. Stocks: 'The same consequences follow from composing the heaven of numbers, as some of the Pythagoreans do who make all nature out of numbers.'

29. Plato, *Republic*, 530d.

30. Archytas Fragment 1.

31. Archytas Fragment 4.

32. Simplicius *Physics Commentary*, 467, 26 ff.

33. See T.L. Heath, *A History of Greek Mathematics*, 2 vols, Clarendon Press, 1921; I. Mueller, Greek arithmetic, geometry and harmonics: Thales to Plato, in C.C.W. Taylor (ed.), *Routledge History of Philosophy Vol. I: From the Beginning to Plato*, Routledge, 2001, pp. 271–322; C. Huffman, *Archytas of Tarentum: Pythagorean, Philosopher and Mathematician King*, Cambridge University Press, 2010.

34. L is in proportion to a, as a is to b, as b is to 2L.

35. See A. Barker, *Greek Musical Writings*, Cambridge University Press, 1989, p. 50; C. Huffman, *Archytas of Tarentum*, pp. 63–4 and 419–20.

36. See Barker, *Greek Musical Writings*; Huffman, *Archytas of Tarentum*.

37. A.E. Taylor, *A Commentary on Plato's Timaeus*, Clarendon Press, 1928.

38. Plato, *Phaedo*, 86b ff.

39. See Simplicius, *Commentary on Aristotle's Physics*, 467, 26.

40. See Aristotle *de Caelo*, 293a18ff., Aetius II, 7,7 ff.

41. See K.R. Popper, *The Open Society and its Enemies*, Routledge and Kegan Paul, 1945, p. 248 note 9 and Popper, The nature of philosophical problems, p. 87ff.; also R. Wright, *Cosmology in Antiquity*, Routledge, 1995, p. 54.

42. Plato, *Timaeus*, 35c.

43. Plato, *Timaeus*, 36d.

44. Plato, *Timaeus*, 35d.

45. Plato, *Timaeus*, 53d4–54a6.

46. Philolaus Fragment 1, cf. Fragment 2, esp. 'it is clear that the cosmos and the things in it were fitted together by limited and unlimited things'.

47. Or indeed whether it fitted itself together in the manner of the Milesians.

48. Philolaus, Fragment 7.

49. Philolaus, Fragment 6.

50. Philolaus, Fragment 4, cf. Fragment 3, esp. 'there will not be anything that is going to know if everything is unlimited'.

51. The *locus classicus* here is Plato's *Timaeus*, see A.D. Gregory and R. Waterfield, *Plato: Timaeus and Critias*, Oxford University Press, 2008, and see also Chapter 4 in this volume on Kepler.

52. J. Kennedy, Plato's forms, pythagorean mathematics, and stichometry, *Apeiron* **43**(1), 2010, 1–31, *The Musical Structure of Plato's Dialogues*, Acumen Publishing, 2011; cf. A.D. Gregory, Kennedy and stichometry: some methodological considerations, *Apeiron*, **45**(2), 157–79.

CHAPTER 3

Divine light

ALLAN CHAPMAN

'Thou deckest thyself with light as it were with a garment', *Psalm* 104:2

(Book of Common Prayer)

Most of the great civilizations of the world have left either religious or philosophical writings about light. Ancient Egyptians saw it as a blessing from Ra, while the Buddha spoke of inner illumination. It was, however, through the classical Greek, Judeo-Christian, and, between AD 800 and 1100, the Islamic traditions that one traces much of our knowledge not only of optics, but also of light-philosophy, as they came to influence the modern world.

In the Judeo-Christian tradition, light was one of the primal agencies of creation, and in *Genesis* 1, God separates it from the tangible darkness to illumine the newly formed world. As the author of *Psalm* 104 (cited above) tells us, God wears light 'as a garment' in a radiant manifestation of His creative majesty. And in the Islamic tradition of science and religion there is a similar parallel, as when in around AD 1000 Ibn al-Haytham, known in the Latin tradition as Alhazen, says 'God is the light of the heavens and the earth'.[1]

It is hardly surprising, therefore, that this image of the power of light should have been so formative in subsequent philosophy, and that it came to be intimately related to mathematics, as personal illumination, logic, and eternal truth all came together. And while light was a thing of visual beauty, and partook of the ineffable, it was also replete with puzzles and paradoxes. Why, for example, did this wonderful radiant force have no tangible substance? How did it travel from its source to the beholding eye, to connect our human perceptions to the wider universe? What was its relation to geometrical truths? And one of the biggest puzzles of all: what was colour and why did it so enchant us; yet where did colour go in the dark?

Light in Greece and Rome

One of the oldest optical problems in Greek thought was how light related both to the human eye and to the psychology of perception. In many respects, this problem ran through Platonic thinking in the fourth century BC (or really the fifth century, if one locates Plato's ideas in the teaching of Socrates). In this way of thinking, seeing was an extension of touch, as the eye emitted some kind of feeler that connected it – and by extension, the sentient soul of the beholder – to the object perceived. This *extramission*, or emitting, theory of light fitted nicely into the Platonic idea of a realm of divine perfect Forms that we could conceive in our minds, yet never create here on the corrupt earth. We could, for example, conceive, or *see*, in our mind's eye a perfect pot, yet no earthly potter could actually make such a pot! For as we needed the light to see objects so, to Plato, we needed an apprehension of 'the Good' to connect with the realm of perfect and eternal Forms.[2]

The problem with extramission ideas of light, however, lay in our inability to see objects in the dark. For surely, if one could feel, touch, taste, and smell something like a lump of incense in the dark, why could one not see it if our eyes emitted seeing rays? It was Aristotle, around 350 BC, who began something of a critique of this extramission theory – discussing the problem of our failure to see in the dark – though it would still be several centuries before an *intromission* explanation of seeing won the ground. In the intromission theory, of course, it is assumed that light exists in its own right in the universe and enters our eyes from outside, and that if the light is blocked off, then our vision fails.

Another source of optical fascination for the Greeks, as it would be for later Arab and European philosophers, was the nature of pinhole images. For why did a pinhole project an image of the brightly lit outside world into a darkened room? It was Aristotle – a believer in the intromission theory of light – who first discussed 'pinhole' images (or more correctly, why the sun, when shining through a crack, projects a perfectly circular solar image) in his *Problemata* (Queries about Nature) of *c.* 350 BC.[3]

This psychological, perceptual, discussion of light naturally related to the problem of colours. Were the colours somehow innate within the objects themselves, and thereby tainted the white light falling upon them, or did the colour exist in the light itself? A question which, on one level, would rumble through the centuries ahead, to become part of the debate about the unchanging *primary* qualities (shape, hardness, and size) and mutable *secondary* qualities, such as colour and taste, about which people may disagree.

The rainbow was to supply an eternal source of fascination both physically and spiritually. God, in the Judeo-Christian tradition, for example, had set the rainbow in the heavens after the subsiding of the waters of the Great Flood to stand as a covenant to Noah that He would never again inundate the earth. Its spiritual source and significance apart, however, most classical philosophers, such as Aristotle and Ptolemy, were in agreement that the rainbow and its colours were brought about by natural means and had some kind of geometrical relationship with the sun.[4]

Pliny the Elder in his *Historia Naturalis* suggests that not only do rainbows always face the sun, but that they are generated by a ray from the sun striking a hollow in a cloud and being refracted or reflected back in a semi-circular form. According to Pliny, the colours were produced by the light being somehow intermixed with 'clouds, fire, and air'.[5] A problem that would be placed on a firmer footing, scientifically speaking, by Alhazen around AD 1000, as we shall see in the next section.

The ancients were on firmer ground, however, when dealing with the geometry of light: where something, at least, could be measured with some accuracy. Euclid, the great systematizer of geometry, around 290 BC, saw light as moving in straight lines with relation to the beholding eye as an implicit part of geometry, as he outlines in the early books of his *Elements*.[6] This was to lie at the heart of the subsequent association of optical studies with ideas of *perspective*, and the real and visual appearances of bodies – a crucial prerequisite, for example, for Vitruvius' *De Architectura* (*c*. 15 BC).

Two pagan figures advanced optical study considerably at the beginning of the Christian era, however – although still working within the explanatory framework of the extramission theory of vision, in which light, or 'seeing', somehow radiated from the eye to touch the object under observation. Hero of Alexandria (AD 10–70) and Ptolemy (*c*. AD 90–168) both did important work on the nature of reflection and refraction, and Ptolemy saw the process of vision not as a simple straight line, but rather as an ever increasing *cone*, the apex of which was seated in the eye: a useful mechanism for explaining perspective.

Unfortunately only parts of Ptolemy's *Optics* have survived, primarily via Arabic intermediaries, but what we do have gives us an insight into his work on reflection and refraction, and how light rays are bent when entering a denser medium, such as when moving from air to glass or water. And as a great systematizer of what had gone before – in the same way as he had incorporated much earlier Greek astronomy into his *Magna Syntaxis* (or *Almagest*) – he includes the work of earlier – and otherwise to us, lost – Greek optical writers in the *Optics*.[7]

And one can understand how optics and vision so fascinated the inquisitively minded ancients, faced as they were with puzzles as diverse as the philosophy of perception, how plane geometry related to perspective, and how pinholes projected images. Cleomedes noted around AD 180 how a coin concealed at the bottom of an empty cup could be made visible by pouring water into the cup and somehow refracting the light.[8]

Yet,why was it the Greeks who came up with these ideas, and not the Egyptians or Babylonians? This, of course, is a very big question, and too complex to be gone into here, but I would suggest that it has something to do with the Greeks being extensive travellers, their concern with political freedom, and the independent money that their essentially trading, mercantile economy facilitated, as opposed to what was possible in the theocratic hierarchies of the Nile, Euphrates, and Tigris valley cultures. An economic, social, and intellectual freedom, indeed, which enabled them to invent not merely philosophical mathematics, but also civic life, public theatre, sporting contests, and 'celebrity' teachers, philosophers, poets, charioteers, and comedians. They were the first 'thinkers', in fact, to come down to us not just as ideal types, but as named idiosyncratic characters in their own right.

Medieval Arabia

The period *c.* AD 850–1300 saw remarkable innovation in science within the Arabic world. Al-Kindi of ninth-century Kufa and Baghdad (and a little later, Ibn al-Haytham) was perhaps the first Islamic optical writer, and his experiments on light rays and transparent spheres led him towards an approximation of what in the seventeenth century would become Snell's Law, from the Dutch researcher Willibrord Snell.[9] Then a century later the Persian Ibn Sahl, in Baghdad, wrote one of the first treatises on lenses, mirrors, and burning glasses, and he, like Al-Kindi, was in part inspired by Arabic translations of Ptolemy's *Optics*.

Like their Greek forebears and later Christians, the Arabic optical writers were fascinated by the spiritual, religious, and philosophical aspects of light, which they saw as a gift of God. And by Arab, one speaks of a culture that extended from Andalusia in southern Spain across north Africa, Cairo, and around the fertile crescent into Persia and even into Mongol Uzbekistan. Where the Arabs were especially significant, however, was in their development of a mathematical and experimental approach to the study of light. They were, moreover, pretty well from Al-Kindi onwards, working within an intromission theory of light, in which light was divinely present in the universe and entered mirrors, lenses, and human eyes from the *outside*, rather than being an emitted 'feeler'.[10]

Ibn al-Haytham, also known as Alhazen, was not only the Islamic world's greatest optical researcher and writer, but was – arguably – the greatest of all the medieval Arabic scientists. Indeed, one is still amazed at his imaginative brilliance, especially as an experimentalist, and by his ability to proceed from the details of a 'laboratory' experiment to draw conclusions about nature in general.

A native of Bosrah, in Syria, Alhazen spent most of his 75-year life working in Cairo, where he variously practised medicine, taught, and even acted as a manuscript copyist as a way of making a living, down to his death in 1040. The nature of refraction in its various forms was a theme running through much of his *Kitab al-Manazir*, or 'Optical Thesaurus' as it would come to be known in twelfth-century European Latin translation (Figure 3.1).[11]

Alhazen's *Kitab* contains one of the earliest detailed descriptions of the eye, its structure, function, and 'seeing': how ophthalmic optics relate to perception and vision. He dissected eyes – presumably those of butchered animals in view of the Koranic prohibition of human dissection. He traced the optic nerve between the brain and the eye, and within the eye itself, gave an elegant description of its working parts: the aqueous and vitreous humours, the lens, iris, and conjunctiva. Alhazen dissected out the flexible lens, and studied how it brought an image to a curved focus within the concave retina. (Indeed, several Arab doctors in the Middle Ages, most notably Albucasis, wrote upon eye disorders, and developed an operation of 'couching for cataract', whereby the skilful insertion of very fine needles into the eye could break off, and even remove, a cataract-clouded lens, at least letting the patient see a bright, albeit blurred world.[12])

Alhazen also experimented with glass or quartz spheres, in water and in air, to study refraction and the appearance of colours, as well as the camera obscura.

One area to which he and his disciple Kamal al-Din al-Farisi made a serious contribution was the optical and geometrical basis of twilight. Probably using an astrolabe, he measured star positions in the pre-dawn sky, and the angular

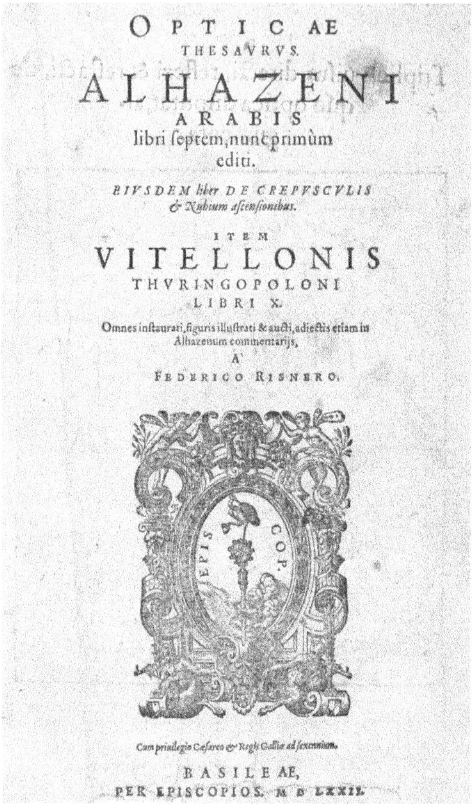
OPTICAE
THESAVRVS.
ALHAZENI
ARABIS
libri ſeptem, nunc primùm
editi.
EIVSDEM liber DE CREPVSCVLIS
& Nubium aſcenſionibus.
ITEM
VITELLONIS
THVRINGOPOLONI
LIBRI X.
Omnes inſtaurati, figuris illuſtrati & aucti, adiectis etiam in
Alhazenum commentarijs,
A
FEDERICO RISNERO.

Cum priuilegio Cæſareo & Regis Galliæ ad ſexennium.
BASILEAE,
PER EPISCOPIOS. M D LXXII.

Figure 3.1 Title page of the Latin translation of Alhazen's *Kitab al-Manazir*, or *Optical Thesaurus* printed by Friedrich Risner in 1572. Image courtesy of Université de Strasbourg, Service Commun de la Documentation.

motion and duration between the first glimmering of dawn in the zenith and the sun's appearance on the horizon. Alhazen calculated that the sun was 19° below the eastern horizon when twilight began. And the cause of twilight, or course, was the refraction of sunlight in the earth's atmosphere.[13]

In the dry climate of north Africa and much of the Arab world rainbows are not especially frequent, but Al-Farisi built upon Alhazen's own atmospheric optical researches and directed a ray of light into a suspended glass sphere in an attempt to determine its path, by internal reflection and refraction, into and out of the suspended sphere.

Alhazen's impact upon the subsequent study of optics was enormous. His *Kitab* was translated into Latin in the twelfth century, and without doubt became the starting-point for much post-classical Western experimental and theoretical optics. And as indicated above, Alhazen, just like his Greek predecessors, saw the study of light as a noble science that enabled human beings to glimpse the mind of God. A recognition of the divine, indeed, that passed through the Greek pagan apprehension of the 'Logos', through mystical Islamic thought, and into the Jewish-Christian tradition of medieval Europe.

Robert Grosseteste of Oxford and Lincoln

A poor boy from rural Suffolk, Robert Grosseteste ('Great Head') obtained his education in the hurly-burly world of the early Oxford, and probably Paris, universities. At the beginning of the thirteenth century, when in his 30s, he was greatly impressed by the teachings of his Italian contemporary, St Francis of Assisi, and joined the young Franciscan Order. A brilliant and inspiring teacher, Robert became Lecturer in Theology to the Oxford Franciscans in 1224 and, like many Franciscans, being fascinated by the divine command 'Let there be light', he came to see light as God's very presence suffusing the universe.

Influenced by the Bible and classical and some Arabic writers, Robert became deeply interested in light, seeing, and the very nature of perception. He became also Western Europe's first great philosopher of science, for while there had been distinguished mathematical European Christians before him, such as the Venerable Bede of Jarrow and the Frenchman Gerbert of Aurillac (later Pope Sylvester II), Robert's thought was much more systematic and encompassing. It was also he who saw number and mathematics as lying at the heart of *scientia* or knowledge: a concept that he developed from Boethius (who was executed in AD 524 for daring to criticize the absolutist tyranny of his master Emperor Theodoric).[14]

To Robert Grosseteste light was the 'First Form' of Creation – apparently combining Biblical and Platonic themes – and two of his books in particular dealt with light. *De Luce* ('On Light') dealt with the metaphysics of light, *De Iride* ('On the Rainbow') discussed the rainbow, while his *De Sphera* also touched upon optical issues.[15]

In addition to light, yet intimately associated with it, were two other ideas that Robert discussed at some length, and which both came to play a major part in subsequent scientific understanding. The first of these was mathematics and geometry, which he saw as underpinning our logical understanding of nature. Once again, his stress on mathematics and number carried ancient classical connotations back to Plato and Pythagoras and, to some extent as a result of

Robert's impact upon unfolding Western thought, would be taken up by the fourteenth-century 'Merton College Geometers' in Oxford, and then by Kepler, Galileo, Newton, and beyond. For long before the end of the Middle Ages astronomy, optics, and the physical sciences were already well down the road towards mathematization.

Second, Grosseteste played a major role in the development of a coherent scientific method, based upon the careful analysis of phenomena, resolving them into parts, and then seeing if they could, intellectually speaking, be re-assembled. This method was intended to enable us to extrapolate unchanging principles, or what would later come to be seen as *laws*; and once these principles had been elucidated, they could be used to make predictions. And hopefully, these in turn should be amenable to testing: important components, indeed, in the evolution of a coherent understanding not only of natural phenomena, but of how those phenomena might relate to the divine and human minds.

Robert Grosseteste's academic and ecclesiastical career proceeded apace in the 1230s by the time he was in his 50s. He became Chancellor of Oxford University, and in 1235 Bishop of Lincoln: the great ecclesiastical see that sprawled across central England and of which Oxford occupied a southern extremity. He died in 1253, in his late 70s: a truly venerable age for the thirteenth century.

Friar Roger Bacon (c. 1214–92), of Oxford and Paris

Alhazen's *Kitab* or 'Optical Thesaurus' was translated into Latin towards the end of the twelfth century, and while it was probably known to Grosseteste, it was Robert's disciple (but maybe not his direct pupil) Roger Bacon, another Oxford Franciscan, who first realized its significance and in particular developed optics further (Figure 3.2).

To Bacon, however, light and optics were part, albeit a major part, of a broader scientific, theological, and philosophical enterprise. And when the already early-middle-aged Bacon became a Franciscan, around 1257, theology, philosophy, and science were not conceived of as distinct disciplines so much as interconnected parts of a wider divine whole. And what Friar Roger wanted to do was develop an all-embracing body of Christian *scientia* that was integrated and logical. Not dissimilar, in fact, to the great pagan system of Aristotle, whose works were beginning to filter into Europe from the Byzantine and Arabic

Figure 3.2 The Franciscan Roger Bacon in the Oxford University Museum of Natural History. He is shown holding an armillary sphere. Photograph courtesy of OUMNH.

worlds and become accessible in Latin translation, and upon whose philosophy the young Bacon was said to have lectured in Paris.

Along with Aristotle, Bacon saw two other figures as models for his 'renewal' of learning: Solomon and Ibn Sina (also known as Avicenna). For Bacon regarded God's truths, including those pertaining to nature and the intellect, as divine revelations in their own right. But after these Jewish, pagan, and Muslim giants, Bacon argued, it was now time for a consummation of learning from a *Christian* perspective.[16]

With Bacon, however, there are several key parts of his life and career about which we are uncertain, especially his movements and activities in and between Oxford, where he studied and sometimes lived, and Paris. Suffice it to say that by the mid 1260s his writings and reputation were being discussed with interest in Rome, and around 1268 he sent a collection of his *opera* to Pope Clement IV in the hope of gaining Papal support for his enterprise. But sadly, owing to His Holiness's death on 29 November 1268, things stalled.

Friar Roger's grand enterprise, however, was intended to examine, amongst other things, mathematics, astronomy, and astrology; natural science and its relation to theology; the use of experimentation; and optics and perspective. This, and his wider scheme for an integrated system of learning, were set out in his three great treatises, *Opus Majus*, *Opus Minus*, and *Opus Tertium*. And within these, *De Perspectiva* was primarily concerned with light and seeing.[17]

Ptolemy and Alhazen were Bacon's principal inspirations in optics, in conjunction with the geometrical certainty of Euclid. And central to his thinking was a rational inquiry into what might be called 'action at a distance': that is, how one thing affected another across space or distance; for example, how did we see the light of the stars, and receive the brilliance and heat of the sun? Indeed, this all formed part of what Bacon called the 'multiplication of species': not 'species' in the biological sense in which we tend to use the word today, but as proliferating forms of properties or even *energy* in nature. How, for instance, do light, heat, colour, magnetism, astrological, and even psychic forces relate to each other? (Astrology, let us not forget, was not in itself deemed superstitious in pre-modern science, as the earth-encircling celestial bodies focused their particular influences on the centre of the creation, to affect the tides, winds, four terrestrial elements, and four 'humours' of the human body.)

Also fundamental to Bacon's thinking was a sense of *naturalism*, for in his scheme of things *natura* possessed its own logic and way of acting, in a manner ordained by God at the creation. This way, moreover, should be accessible to the human intellect, especially via number, mathematics, and reason. For many things attributed to magical or demonic forces by the ignorant were really parts of natural operation still waiting to be fathomed by humans. A way of thinking, indeed, with clear resonances extending back to Grosseteste, Al-Kindi, and beyond.

And while it is all too easy, with the gift of hindsight, to read 'modern' scientific ideas and methods into Bacon, what cannot be denied is that he was in many ways what later ages would style an 'experimentalist', in so far as he saw the value of examining, measuring, and testing nature as a preliminary to trying to understand natural phenomena.

Crucial to our understanding of light in medieval Europe, however, was the Polish-Thuringian scholar Witelo or, as he was known in the Latin, Vitello, whose *Perspectiva* was probably completed around 1278. Influenced by Alhazen's *Kitab*, he was interested in the psychology and spirituality of perception, seeing light as divine and a component of the First Cause of Creation, as well as in the more technical aspects of refraction and reflection.

Yet it was in the study of the rainbow that medieval optics came into its own.

Bacon, Theodoric, and the rainbow

One of the great research projects of the medieval centuries, in both Arabia and Europe, was the behaviour of light in the earth's atmosphere. This had roots going back to Aristotle's *Meteorologica* (*c.* 350 BC), Ptolemy, and then Al-Kindi, Al-Farisi, and Alhazen who, by *c.* 1020, had produced a coherent and mathematically and observationally based model for twilight. But it was in the damp, rainy climate of northern Europe that studies of the rainbow, mock suns, and solar and lunar 'haloes' and 'glories' would be significantly advanced.[18]

Robert Grosseteste had discussed such phenomena, especially in *De Iride*, but it was with Roger Bacon that the understanding of optical geometry really began to move forward. By measuring – presumably with an astrolabe – the angle of the sun in the sky, along with the height of the bow, Bacon concluded that coloured light exited the rainbow at a consistent angle of 42° from the incident ray going into the cloud. He built upon earlier ideas that the rainbow was part of an arc which, if continued under the earth, would form a circle which in itself was the base of a cone. The apex of the cone would be the sun, and the observer stood on a line connecting the apex to the centre of the base circle.[19]

Thus, the biggest rainbows would occur soon after dawn or before sunset, when the sun was low in the sky, and the bow high, with the observer directly facing the cloud and the sun behind. Friar Roger also mentions 'artificial' rainbows generated from spray, such as from a water-wheel, when the sun, the observer, and the spray were in the correct positions. And like Alhazen, Roger experimented with glass spheres and lenses, about which more will be said anon. Bacon appears, however, to have seen the bow as produced by the *cloud* as an entity, rather than by the billions of individual water droplets which make up the cloud. That next realization would be advanced in the next generation through the researches of Theodoric (or Dietrich) of Freiburg, who was active between *c.* 1290 and *c.* 1310.

Like Roger Bacon, Theodoric (as I shall refer to him) was a Friar – a member of that other great early thirteenth-century mendicant and teaching order, the brethren of St Dominic. Indeed, it is fascinating how the Franciscan and Dominican mendicant Friars thought it natural to combine a life of poverty, chastity, and Christian service to the poor with the most advanced theological thinking of the age and an active involvement in what we would now call scientific research. (Three centuries later, they would be joined in this compound quest by the newly founded Society of Jesus, and all three Orders attracted, and continue to attract, some of the sharpest and most adventurous intellects within the Roman Catholic world.)

Theodoric moved from thinking of the rainbow as occasioned by a simple cloud to being the product of a seeming infinity of individual water droplets, with their own individual refractions and reflections. Building on the earlier work of Bacon and the Arabic opticians, Theodoric became a pioneer of what might be called the 'experimental environment' in which to analyse very specific components of nature, rather than attempting to draw conclusions piecemeal from the entirety.

Theodoric stressed the importance of *individual* droplets of water in the formation of the rainbow. For this, he employed two experimental tools: a crystal ball and a physician's spherical urinal glass (used by doctors to examine a patient's urine) filled with clear water. He shone rays of light into the spheres, which enabled him to ray-trace the internal reflections and refractions within them and establish the exit angle of the now colour-fringed light. Following the Arab writer Ibn Rushd, or Averroes, he explained the coloured light as being formed from varying combinations of light and dark.

Theodoric recognized that colour was related to refraction, pointing out that when a ray passed straight through the spherical vessel it came out white, yet when it was refracted the colours appeared. He also addressed the question of why the colours of the spectrum became wider and more conspicuous the further the light ray travelled from the sphere: this was caused by the pure white somehow absorbing, or becoming contaminated with, the ambient darkness through which it was passing. A 'corrupted whiteness' theory of colour, indeed. Red light, however, was always found to be at an angle closest to the original path of the incident ray. (In short, red light seemed least bent, or refracted, than any of the other spectral colours.[20])

By *c.* 1310 Theodoric had taken experimental optics about as far as they would go before the seventeenth century. And by then, not only was there an experimentally based, and largely correct, model for the primary and reversed secondary bows of the rainbow based on droplet refraction and reflection, but the basics of its conical geometry were there as well. And very importantly, Theodoric had shown how experiments conducted on transparent spheres and single rays of light might yield significant insights into large-scale natural phenomena, a question much argued over in classical and medieval times. For how could a 'microcosm' experiment in the laboratory, conducted with human hands and tools, really reveal the underlying truths of God's creation? A debate, in fact, which would continue on until the seventeenth century.

And this is how things would stand until the time of Robert Hooke and Sir Isaac Newton in the second half of the seventeenth century. What would continue in the interim, however, would be a growing fascination with optical devices.

Light in action: early optical instruments

Magnifying lenses, concave mirrors, burning glasses, and other such devices were well known by the death of Bacon in the early 1290s. But it was allegedly Friar Roger who first suggested the use of lenses held before the eyes as an aid to defective vision. For, in the translation by Robert Smith of the passage from *Opus Majus* in 1738, 'If the letters of a book . . . be viewed through a lesser segment of a sphere of glass or crystal, whose plane base is laid upon them, they will appear far better and larger.' In short, with a plano-convex lens laid flat-side-down above a page![21]

Various suggestions have been put forward concerning who invented spectacles, or 'visual glasses' proper, and when. Tradition cites two Italians – the Dominican Friar Alessandro della Spina and his friend Salvino D'Armate – on the evidence of D'Armate's 1317 Florentine tombstone, where he is styled 'Inventor of Spectacles'. On the other hand, other claims have pushed the invention date as far back as 1285.[22]

What is significant, however, is that something so useful as spectacles soon caught on, and by the early fifteenth century they had become the trademark of the scholar in many Western European art works. Pictures of St Mark (author of the earliest Gospel) and of St Jerome (the *c.* AD 400 Latin 'Vulgate' Bible translator), for instance, often show them with a pair of glasses (Figure 3.3). I also possess a postcard of a German manuscript with a semi-cartoon depiction of a *c.* fifteenth-century schoolmaster: there he sits, robed and biretta'd, on his high chair, birch in hand, and with a pair of glasses tied around his head. And when El Greco came to paint the portrait of Cardinal Don Fernando Niño de Guevara in Spain, *c.* 1600–1604, his Eminence is shown sporting a pair of thin wire-framed glasses of the type that John Lennon turned into a fashion accessory in the 1960s (Figure 3.4).[23]

Figure 3.3 St Mark, in an engraving by Lucas van Leyden (1518), is represented with spectacles to highlight his scholarly role as the author of the earliest Gospel. He is frequently associated with a winged lion in Christian iconography.

With spectacles becoming so relatively commonplace by the beginning of the sixteenth century, one can only assume that the art of the lens grinder must have been fairly widespread across Europe, at least in the big cities. And there were enough spectacle makers in the City of London by 1629 for them to found their own livery company.

But if spectacles represent the first application of optics to everyday life, what evidence is there for a 'pre-Galileo' telescope? Once again, there are some rather interesting, albeit highly ambiguous, remarks in Roger Bacon, such as the mention of an optical arrangement enabling him to see 'from an incredible distance . . . the smallest letters, and may number the smallest particles of dust and sand . . .'.[24]

Figure 3.4 El Greco's portrait of Cardinal Don Fernando Niño de Guevara in Spain, c. 1600–04. His Eminence is shown proudly sporting a pair of glasses. Image courtesy of the Metropolitan Museum of Art.

Then there was the supposed 'Tudor telescope', or an arrangement of convex and concave mirrors mentioned by Leonard Digges in 1571 whereby one could clearly see the streets of a distant village, including small objects.[25] Indeed, Leonard Digges and his son Thomas were fully familiar with the optical devices of Friar Bacon, or at least what had come down from or was understood about Bacon, either correctly or apocryphally, when Thomas was re-publishing his deceased father's works in 1571. A few years later, however, in an undated manuscript in the Lansdowne MS in the British Library, is an account of a configuration of concave and convex mirrors that could magnify distant scenes, which the instrument maker William Bourne supplied to Lord Burghley, Queen Elizabeth I's chief advisor of state.[26]

In the early 1990s, moreover, Colin Ronan and Gilbert Satterthwaite, both now sadly deceased, attempted a reconstruction of the Digges/Bourne instrument, which was demonstrated in the rooms of the Royal Astronomical Society at Burlington House, London. As I knew both gentlemen very well, I was invited to observe a distant object through what was really a 'Galilean'-type optical system that used mirrors rather than lenses. It certainly produced a small-field, aberrated, yet undoubtedly magnified image.

This *may* – and it is a very big *may* – have been similar to the 'perspective glasse whereby was shewd manie strange sightes' to the native people of Roanoke, Virginia, over 1585–6, by Thomas Harriot on the Ralegh–Grenville expedition to North America, as later published by Harriot in 1588. And Harriot, one must recall, was a member of the Digges, John Dee, and Sir Walter Ralegh circle in London.

On the other hand, the 'strange sightes' shown to the Algonkian tribal elders need not have been telescopic, for by his own confession Harriot also showed the Indians burning glasses, distorting mirrors, and other optical novelties which would certainly have qualified as 'strange sightes'. Even so, Thomas Harriot's lecture-demonstration of a variety of optical, mechanical, and magnetic devices probably qualifies as the first ever 'modern science' lecture to have been delivered on the North American continent – and that to an indigenous audience. An event, indeed, deserving to be commemorated in modern North American scientific and educational culture![27]

The 'Dutch Truncke'

It is my suspicion, however, that whatever Thomas Harriot showed to the Roanoke locals in 1585–6 did not produce a useful telescopic image, for while Harriot made no claims to having invented the telescope, he was the first to use a post-1608 two-lens refractor to view an astronomical body, and to leave a documented account of what he saw.

At 9.00 p.m. on the evening of 26 July 1609, when he first pointed his newly-acquired 'Dutch Truncke' at the five-day moon, Thomas Harriot was a 49-year-old well-to-do and somewhat reclusive bachelor mathematician. Indeed, a man at the forefront of European mathematics, who corresponded with Kepler and other major Continental mathematicians, enjoying a residence in the grounds of Syon Park, near London, the seat of his very good friend Henry Percy, Ninth 'Wizard' Earl of Northumberland – Lord Percy currently being in the Tower of London having been rather distantly implicated in the Gunpowder Plot of 5 November 1605.

Harriot's account of the telescopic moon on the evening of 26 July is still preserved in his sketch, along with numerous subsequent moon maps, sunspot and Jovian moon drawings, in the West Sussex Public Record Office, Chichester, where I have examined them.[28]

What concerns us here, however, is the obvious amazement and delight which Harriot and his friend Sir William Lower expressed at their first views of the telescopic moon. Let us remember that Harriot in July 1609 was observing the heavens *four months* earlier than Galileo; yet when he learned of Galileo's discoveries, probably by letter from Kepler in the early summer of 1610, before he had acquired a copy of Galileo's *Sidereus Nuncius* (published in Venice, March 1610), Harriot expressed only *delight*.

This all suggests, I would argue, that not only was Harriot singularly unconcerned with discovery priority (a thing that often irritated his friends); but that, whatever the 'Tudor telescope' arrangement of the 1570s may have been, it did not compare in optical potential with the new 'trunckes', 'perspectives', or 'cylinders' that were coming out of Holland in 1609. In short, the lens-based refracting telescope of the pattern with which Galileo was to make his discoveries was a very different instrument indeed from that of Bacon, Digges, Bourne, and the young Harriot.[29]

Understanding God's 'pallat': new ideas about coloured light, 1604–1704

The surviving works of the Tudor telescopic and optical writers were not really concerned with religious questions, being primarily about practical technology and experimentation. In the seventeenth century, however, several major scientific men addressed themselves to the problems of understanding light, all of whom left remarks about its relation to the divine. Johannes Kepler's *Astronomiae Pars Optica* (1604) really took up where Alhazen, Bacon, Witelo, and Theodoric had left off, both in his analyses of the physiological structure of the human eye, and in studies of the geometrical behaviour of light itself and the formation of retinal images.

One of the most significant insights he had, however, was that the intensity of light diminished in an inverse square mathematical progression the further away from the light the observer moved. A proportionate relationship, indeed, which would have profound implications for other branches of physics, such as gravitation. And the devoutly Christian Kepler would also be concerned with how the *soul* of the beholder saw and perceived when observing the Divine Light.

Yet coloured light, as opposed to optical geometry, had not really been Kepler's concern, although he did discuss the newly discovered telescopic images in his *Dioptrice* of 1611, in which he proposed the use of the subsequently named 'Keplerian eyepiece' for telescopes. This eyepiece made a more efficient use of the light refracted by the telescope's object glass, resulting in a bigger, brighter, and less aberrated image than did the 'Galilean' eyepiece.

René Descartes too was fascinated by light: how we receive it physiologically to produce retinal images and how we might improve optical instruments, and did a series of brilliant mathematical analyses of the rainbow. And while his *Dioptrique* (1637) is his best-known work on optics, a fascination with light and optical geometry runs through several of his works.

And as with all other natural phenomena, Descartes saw light as the product of *motion*, for in the Cartesian philosophy structured movements within and between vortex systems lay at the heart of physics – and all natural science, at bottom, was physics. Light for Descartes, moreover (as it was for most of his contemporaries), was instantaneous in its motion: for if the whole of creation was but a vast machine of moving 'corpuscles' or particles that pushed each other, then there really was no room between the corpuscles to allow the necessary knock-on effect that would be necessary to produce a *temporal* transmission of a light-ray from a star to the observer's eye. Rather, the mechanical pushing of the corpuscles was so tight as to resemble something like the passage of a mechanical impact down a solid rod. Hence, light *must* be instantaneous in its travel!

It was the angles of impact of the light rays, and their relationship to the retina, that produced colour in Descartes' philosophy. And while there has been much debate among philosophers about the exact nature of Descartes' personal religious beliefs, the impression comes across in his writings that he saw the source of light as divine.

It was two Englishmen, and Fellows of the Royal Society, who came to engage with the nature of colour face-on in the 1660s and 1670s: Robert Hooke and the young (Sir) Isaac Newton. In Observations IX and X in particular, and elsewhere in *Micrographia* (1665), Hooke describes a brilliant series of optical investigations, partly stimulated by the way that star images seem more blue when high in the sky and more red on the horizon. Of course, there

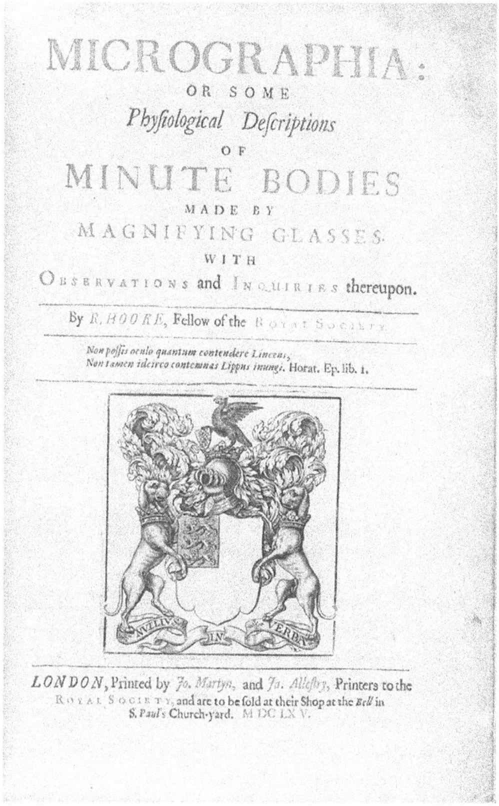
MICROGRAPHIA:
OR SOME
Physiological Descriptions
OF
MINUTE BODIES
MADE BY
MAGNIFYING GLASSES.
WITH
OBSERVATIONS and INQUIRIES thereupon.

By R. HOOKE, Fellow of the ROYAL SOCIETY.

Non possis oculo quantum contendere Linceus,
Non tamen idcirco contemnas Lippus inungi. Horat. Ep. lib. 1.

NVLLIVS IN VERBA

LONDON, Printed by *Jo. Martyn*, and *Ja. Allestry*, Printers to the ROYAL SOCIETY, and are to be sold at their Shop at the *Bell* in S. *Paul's* Church-yard. M DC LX V.

Figure 3.5 Title page of Robert Hooke's *Micrographia*. The book, despite the hefty 1665 price tag of 30 shillings, was an instant success, with Samuel Pepys declaring it 'the most ingenious book I ever read in my life'.

was nothing new in this piece of knowledge; but what matters is the direction in which Hooke – perhaps the most inspired and imaginative experimental scientist of the seventeenth century – took his researches thereafter (Figures 3.5 and 3.6).

The sensation of colour, Hooke argued, was caused by the refraction, or 'Inflexion', of light. He came to this conclusion in part by taking a two-foot-long conical glass flask, and filling it with clear water – not that easy, as he reminds us – in the City of London in 1665. Sunlight was then made to pass through a small hole and enter the water. Hooke found that when he tilted the flask so that the 'inflected' incident ray of light passed in an oblique line, it produced lovely red fringes on a white screen at the bottom. Yet when he increased the tilt of the flask so that the light struck the water at a more acute angle, then

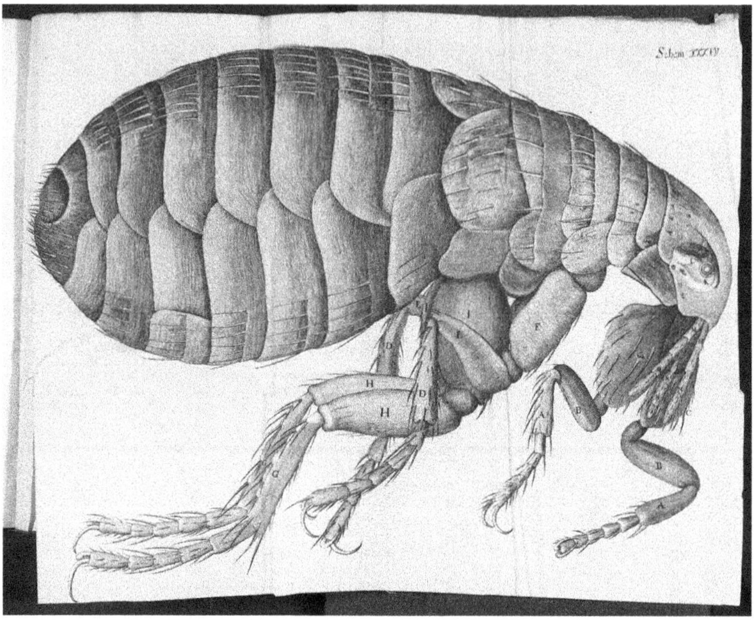

Figure 3.6 An example of the beautiful illustrations in Hooke's *Micrographia*. Observation 53 *Of a Flea* states, 'For its strength, the *Microscope* is able to make no greater discoveries of it then the naked eye, but onely the curious contrivance of its leggs and joints . . . But, as for the beauty of it, the *Microscope* manifests it to be all over adorn'd with a curiously polish'd suit of sable Armour, neatly jointed, and beset with multitudes of sharp pinns, shap'd almost like Porcupine's Quills, or bright conical Steel-bodkins; the head is on either side beautify'd with a quick and round black eye'. Image courtesy of the Bodleian Library, Oxford.

it produced blue fringes at the bottom on the opposite side of the screen. By varying the angle between the extremes, he was able to detect intermediary colours.[30]

So had God painted His creation with a 'pallat' of red and blue primary colours, in a fashion related to that whereby the artist could produce all the colours for his paintings from a skilful combination of primary tints? And Hooke was in a good position to know about practical art, having served for a time in the studio of the great painter Sir Peter Lely before going up to Westminster School and then to Oxford.

Robert Hooke next had made a pair of hollow glass prisms, or 'cake-slice' wedges, that he could fill respectively with red and blue dyes. Adjusting them carefully with two rays of sunlight coming into a darkened room, he was able to produce some interesting colour mixes when directing the ray from each prism on to a screen.

Hooke commented that his researches had opened up 'A Large Window . . . into the Shop of Nature'!

Then he hit a problem. Why was it that when he slid the two wedge-shaped prisms horizontally alongside each other, sharp edges first, while looking through them at a bright light, he saw not colours but only *blackness*? (Or what modern optical physicists would see as a neutral density filter.)[31]

It was this very problem with the two glass wedges that puzzled Sir Isaac Newton, as he wrote in 1672, and helped to spur on his own original optical researches after he had read of it in what he referred to as Hooke's '*Micrography*'.[32]

Yet Hooke was perhaps the first truly original post-medieval researcher to address himself primarily to the nature of colour itself, rejecting Descartes' model of colour generation – as occasioned by 'a certaine *rotation* of the *Globuli aetheri*' – presented in his *Meteorum* (Meteorology *or* Things in the air), Chapter 8 Section 5, on experimental grounds,[33] while referring to his friend Robert Boyle's work on colour. In *Micrographia* Observation IX, where he discussed the water and two-foot flask experiment, Hooke also reported on his investigations of the colours seen in slips of mica – what has come to be called 'Newton's rings' – and contradicted Descartes' mechanical model for colour as generated from rotating globules. And elsewhere he examined the colours seen reflected in silks and studied the anatomy of the eye, the retina, and the nerves. His investigations were extended to astronomical bodies, and in his *Cometa* (1677) Hooke drew attention to the uniform luminosity of cometary nuclei when examined with the telescope. Why, for instance, did that part of the nucleus facing away from the sun glow just as brightly as the part facing towards the sun?

Hooke then proposed his own model for light: it was a pulse, or a sinusoidal wave, impacting upon oblique surfaces, creating the sensation of redness and blueness, with the intermediary colours occasioned by the area between. Indeed, 'the *Phantasm* of Colour is caus'd by the sensation of the *oblique* or uneven pulse of light',[34] perhaps with the blue and red generated by successions of strong and weak pulses.

And like the long tradition of natural philosophers before him, Hooke was unequivocal about the divine source and character of light: a point which he hammered home to the Royal Society in his January–February lectures on light, posthumously published in 1705. Here, indeed, he cites Moses, and the Biblical idea of light as an inextricable part of the wonder of God's creation. He also saw the process of physiological perception, of which light was a major component, as intimately associated with the immortal soul, with knowledge, and with memory.[35]

It is hard to see Newton's fascination with 'the celebrated *Phaenomena* of Colours' without reference to Hooke's work, for *Micrographia* had been published about 18 months before Newton began to examine the 'Phaenomena' himself, and became intrigued by Hooke's 'dark light' wedges experiment in *Micrography*. Of course, Newton's work on colour is too well known to require detailed description here, though there are some points which are especially important.[36]

First, there is Newton's amazingly dangerous experiment of inserting a bodkin into his eye socket, to discover that, by applying pressure upon his eyeball, he could generate colours.[37] Second, there is his famous *Experimentum Crucis*, whereby he found that each of the long-familiar spectral colours produced when light is shone through a glass prism is a physical absolute in its own right and cannot be broken down by a second prism, and that all the colours can be brought together again to produce white light. Indeed, this experiment is one of the great pieces of inspired imagination in scientific history, for it showed once and for all that colour was not a form of corrupted white light, but a real thing in nature.[38]

Third, there was the genesis of that dispute with Hooke which would become the source of much bitterness on Newton's part and would create lifelong enmity between the two men. And as Newton would outlive Hooke by 24 years, it would lead to Hooke's own brilliant and imaginative experimental science being largely side-lined and buried. For Hooke was unable to replicate this feat of re-assembling the first spectrum to produce white light. And quite apart from any prejudice that Hooke may have felt against Newton's rival theory of colour, presented to the Royal Society in 1671, I can say from personal experience that it is *not* an easy experiment to perform. Yes, one can pass each individual colour through the second prism to demonstrate colour's irreducibility, but putting it back together again is very hard, and by no means obvious if you do not know for certain in advance that it can be done.

I personally have only been able to 're-assemble' the colours within the controlled environment of an optical bench, and how Newton did it from a *moving* solar ray through a hole bored in his window shutter in Trinity College, Cambridge, at a 12- or a 22-foot projection distance, I cannot imagine.[39] But he did. And with it, he took the science of optics to a new level. Yet Newton's corpuscular model for light was later shown not to be correct, and Hooke's wave model only partially so. For nineteenth-century optical physicists, most notably Thomas Young, demonstrated that while light does indeed travel in waves, the colour of the resulting ray is determined not by different geometrical parts of one generic light ray, but by the different amplitude frequencies of similar but geometrically distinct waves.

In the twenty-first century, the Newtonian corpuscular theory was combined, as it were, with Hooke's waves, as *photons* were found to have their own vibrative frequencies.[40]

And one has only to read the concluding paragraph of his *Opticks* (1704) to see the centrality of religious inspiration behind Newton's own study, not just of light, but of the whole of science:

> For so far as we can know by natural Philosophy what is the first Cause, what Power he has over us, and what Benefits we receive from him, so far our Duty towards him, as well as towards one another, will appear to us by the Light of Nature.[41]

Notes and references

1. Seyyed Hossein Nasr, *Islamic Science. An Illustrated Study*, World of Islam Festival Publishing Co. Ltd, 1976, p. 142.
2. Plato discusses 'vision' or perception in several *Dialogues*. See *Republic* Pt. III 'Philosopher King', Chapter XXIII, Section VI, in F.M. Cornford (ed.), *The Republic of Plato*, Clarendon Press, 1966, pp. 213–16. Also Paul Friedländer, *Plato, Three Dialogues, Second and Third Periods*, Routledge and Kegan Paul, 1969, pp. 361, 373–6, etc.
3. Aristotle, *Problemata* trans. and ed. E.S. Forster, in *The Works of Aristotle*, Vol. VII Clarendon Press, 1927: see Book XV, Section 6, pp. 911b and 912b.
4. *Genesis* 9:14–17. It was Aristotle who left one of the first optical accounts of the rainbow in *Meteorologica* Book III: *The Works of Aristotle* Vol. III, trans. E.W. Webster, Clarendon Press, 1931, Book III, Sections 4–6, pp. 373–7.
5. Gaius Secundus Pliny [Pliny the Elder], Book II 'Astronomy': *Natural History* trans. John F. Healy, Penguin Classics, 1991, p. 29. Mark E. Smith (ed.), Alhazen's theory of visual perception *Transactions of the American Philosophical Society*, **91**, Parts 4–5, 2001. See Alhazen's *De Aspectibus* Cpt. 7, 6:65–6:68 for his discussion of the refractive properties of 'glacialis' or transparent spheres.
6. Euclid's *Elements* was first published in a printed edition in 1482. I am familiar with the Basel edition of 1533 in the Library of Wadham College, Oxford (text available on Wadham College Virtual Library website). There are easily accessible modern translations of Euclid and Vitruvius.
7. Mark A. Smith, Ptolemy's theory of visual perception: an English translation of the optics, *Transactions of the American Philosophical Society*, **86**, Part 2, 1996.
8. Cleomedes cited with reference in article 'Light', *Encyclopaedia Britannica*, 9th edn, Vol. 14, 1882, p. 577. Most likely from Cleomedes' *De Motu Circulari Corporum Caelestium*, deriving from Posidonius, and containing a variety of optical observations.

9. R. Rashed, A pioneer in anaclastics: Ibn Sahl on burning mirrors and lenses, *Isis*, **81**, 1990, 464–91.

10. David C. Lindberg, *Theories of Vision from al-Kindi to Kepler*, University of Chicago Press, 1976, p. 19. See also D.C. Lindberg, Alkindi's critique of Euclid's theory of vision, *Isis*, **62**(4), 1971, 469–89.

11. Alhazen [Ibn al-Haytham], *Opticae Thesaurus Alhazen Arabis*, Basilae, 1572. Books V, VI, and VII discuss refraction. Book V, Sections 44 and 45 [p. 157] deals in particular with the refractive properties of transparent cylinders. See also Smith, Ptolemy's theory of visual perception.

12. Albucasis, *On Surgery and Instruments*, trans. with commentary by M.S. Spink and G.L. Lewis, Wellcome Institute, 1973, pp. 252–6 for cataract.

13. Alhazen, *Opticae Thesaurus*. See pp. 283–8 for Alhazen's *De Crepulis* ('On Twilight'). '19 grad. quonià est depression solis' ('when the Sun is depressed 19 degrees below the horizon'), p. 288.

14. James Hannam, *God's Philosophers. How the Medieval World Laid the Foundations of Modern Science*, Icon Books, 2009. Many references to Boethius, Bede, and Gerbert. Also A.C. Crombie, *Augustine to Galileo, 1. Science in the Middle Ages, 5th to 13th Centuries*, Falcon Press, 1952; Peregrine, 1969, for Bede and Boethius.

15. A.C. Crombie, *Robert Grosseteste and the Origins of Experimental Science*, Clarendon Press, 1953, Section VI on Grosseteste's 'Metaphysics of Light', pp. 128–34, where he also discusses St Augustine and spiritual illumination.

16. This is developed in the *Opus Majus*. See John Bridges' edition, Clarendon Press, 2 vols, 1897, Vol. 1, Books I–IV.

17. *Opus Majus*, Book V begins Bacon's discussion of optics and perspective. See Bridges *Opus Majus*, Vol. 2.

18. Crombie, *Grosseteste*, Chapter VI, 128–34. See also Aristotle's *Meteorologica* Book III, for the rainbow and mock suns. Nasr, *Islamic Science*, pp. 140–2.

19. Bridges, *Opus Majus*, Vol. 2, Book VI, p. 179. Crombie, *Grossteste*, Chapters IX, 233–59, and X, 260–89. Hannam, *God's Philosophers*, pp. 135–51, for Bacon et al.

20. Theodoric discusses his optical research in *De Iride et radialibus impressionibus* (*c.* 1310), or 'On the Rainbow and impressions created by radiance'. See W.A. Wallace, *The Scientific Methodology of Theodoric of Freiburg. A Case Study of the Relationship between Science and Philosophy, Studia Friburgensia*, The University Press, 1959. Also Crombie, *Grosseteste*, Chapter IX. Edward Grant, *A Source Book in Medieval Science*, Harvard University Press, 1974, p. 864.

21. Robert Smith, *A Compleat System of Opticks*, Cambridge, 1738. Smith translates from Samuel Jebb's *Roger Bacon's Opus Majus*, London, 1733, p. 352. See also H.C. King, *The History of the Telescope*, Charles Griffin, 1955, p. 27.

22. King, *The History of the Telescope*, p. 27. For a sixteenth-century account of how to manufacture lenses and optical mirrors, as well as a variety of optical novelties, see

Giovanni Battista della Porta, *Magia Naturalis* (1558), English trans. John Baptista Porta, *Natural Magick*, London, 1658, Chapter 17, 'Of Strange Glasses', and p. 379 for lens-making.

23. El Greco's painting is described by the holder, the Metropolitan Museum of Art, New York, as 'Portrait of a Cardinal, *probably* Cardinal Don Fernando Niño de Guevara'.
24. Jebb, *Opus Majus*, p. 357. Also cited in Robert Grant, *History of Physical Astronomy*, London, 1852, pp. 515–16. John Bridges in his Oxford 1897 edition of Bacon's *Opus Majus*, Vol. II, Pt. V, 159, also expresses doubt about a 'Baconian' telescopic device: 'But of the combination of two lenses necessary for the construction of the telescope there is no evidence whatever.' Likewise Robert Smith, *Opticks*: see 'Remarks' at end of volume, pp. 111–21, 'I conclude that the time of the invention of the telescope was not earlier than the beginning of the seventeenth century' (p. 121).
25. Leonard and Thomas Digges, *Pantometria*, 1571, Chapter 21. See also Antonia McLean, *Humanism and the Rise of Science in Tudor England*, N. Watson, 1972, p. 149.
26. King, *The History of the Telescope*, pp. 29–30.
27. Thomas Harriot, *A Briefe and True Report of the New Found Land in Virginia* (1588), ed. Henry Stevens, London, 1900, p. 57.
28. The sheet of paper bearing Harriot's 26 July 1609 moon drawing is included in the Harriot Papers in the West Sussex County Record Office, Chichester (Lord Egremont Collection, Petworth House Archives, HMC 241/9 fol. 26); also available on the internet under 'Thomas Harriot moon images'. Reproduced in A. Chapman, A new perceived reality: Thomas Harriot's moon maps, *Astronomy and Geophysics*, **50**, 2009, 1:27–1:33.
29. Sections of Harriot's correspondence about early telescopic optics are included as 'Account of Harriot's astronomical papers' in Stephen Peter Rigaud (ed.), *Miscellaneous Works of James Bradley*, Oxford, 1832. A detailed Harriot optical and astronomical bibliography can be found in Chapman, A new perceived reality and A Chapman, The astronomical works of Thomas Harriot, *Quarterly Journal of the Royal Astronomical Society*, **36**, 1995, 97–107.
30. Robert Hooke, *Micrographia, or Some Physiological Descriptions of Minute Bodies Made by Magnifying Glasses*, London, 1665, Observation IV, 58ff.; also Observation LVIII, 217ff.
31. Hooke, *Micrographia*, Observation X, 74–5.
32. Isaac Newton, A Letter of Mr Isaac Newton . . . containing his New Theory about Light and Colors . . ., *Philosophical Transactions*, **80**(6), 1671–2, 3075–87. See 3084, para. 12, for 'Micrography'.
33. Hooke provides experimental challenges to Descartes' mechanical model of light in several of the Chapters in *Micrographia*. But see in particular Observation IX, 58–61.
34. Hooke, *Micrographia*, Observation X, 67ff.

35. Hooke, in his 'Lectures of Light' delivered before the Royal Society, 1680–2, further took issue with Cartesian theories. They are printed in *The Posthumous Works of Robert Hooke*, London, 1705. The 'Light' lectures are printed on pp. 71–150. See 96–7 for Biblical references.

36. Newton, A Letter of Mr Isaac Newton, 3084, para. 12.

37. Cited from Newton in Richard S. Westfall, *The Life of Sir Isaac Newton*, Cambridge University Press, 1993. This incident is powerfully re-enacted, with full scholarly detail, in Craig Baxter's *Let Newton be* (2009). For full text see *The Isaac Newton Guidebook*, ed. Denis Alexander and Craig Baxter, Faraday Institute, Cambridge, 2012, pp. 121–7.

38. Newton, A Letter of Mr Isaac Newton, 3078, for 'Experimentum Crucis'. Newton also discussed those fundamentally important prism and colour researches some 40 years later in *Opticks* (1704), Book I, Pt. 1.

39. Newton, A Letter of Mr Isaac Newton, 3077, for 22-foot projection distance. In *Opticks*, Book I, Pt. 1, Experiment 6, he also speaks of using a 12-foot projection distance.

40. Thomas Young's research and elegant demonstration of the wave ('undulatory') theory of light - 1800–04 - is discussed in Mary Somerville, *The Connexion of the Physical Sciences*, 3rd edn, John Murray, 1836, pp. 186–8.

41. Newton, *Opticks*, Book III, Pt. 1, Question 31, last paragraph.

CHAPTER 4

Kepler and his Trinitarian cosmology

OWEN GINGERICH

In 1543, the year in which he died at age 70, Nicolaus Copernicus saw his *De revolutionibus orbium coelestium* published. It was a formidable treatise that challenged the traditional view that the heavens – sun, moon, planets, and stars – all revolved around a fixed earth. Today, few of us have any problem endorsing Copernicus' revolutionary proposal of a sun-centred cosmos. Rather, we are baffled as to why it took educated people so long to adopt a heliocentric viewpoint. Was not the ancient Ptolemaic system so cluttered up with ad hoc fixes that it was ready to collapse of its own weight? And despite all the circles on circles, was it not failing to predict where the planets were?[1]

The answer to the second question is that hardly anybody noticed, and as a consequence, no one was cluttering up the Ptolemaic system with fixes. And so we must face a more interesting question. What caused Copernicus to propose such a radical reworking of the ageing cosmology? Where was the Ptolemaic system failing?

Curiously, it was because of the questions the ancient geocentric system did not answer. The sun and moon always pursue eastward courses against the starry background, but the star-like planets are odd. Every year or so they slow down to a stop, move westward for a while, and then take up their eastward course once again. Not only that, but this retrograde movement always took place when the planet was oriented in line with the sun. Why?

When Copernicus threw the earth into motion, the superior planets (Mars, Jupiter, and Saturn) appeared to move backward whenever the swifter moving earth bypassed them. Here was a sensible, geometric explanation of what was happening. What was once a 'fact-in-itself' now became a reasoned fact. Furthermore, with the new heliocentric arrangement, all the planets were organized into an integrated system, and all the planets fell into order according to their periods. Swift Mercury, closest to the sun, had the shortest period, 88 days, while lethargic Saturn, at the outer fringes of the known planetary system, took three decades to revolve around the sun. Here was the beautiful solution to a previously unasked question. It was so elegant that once he found it, Copernicus could not give it up.

Two generations later, with the Copernican system rarely accepted as physical reality, a young German theology student posed a similar conundrum. Why, the young Johannes Kepler asked, are there six planets and what determines their spacing? To this day these are unanswered questions (but in this era of exoplanets around distant stars, the question no longer seems so unreasonable); in any event Johannes thought he had found an answer. There are precisely five Platonic solids – tetrahedron, cube, octahedron, dodecahedron, and icosahedron – just exactly the right number to provide spacers between the six Copernican planets. And they seemed to fit just right (Figure 4.1).

Young Kepler had been a theology student at the University of Tübingen in southern Germany. There the astronomy professor, Michael Maestlin, befriended him and saw his potential in mathematics. Kepler was expecting to become a Lutheran pastor, and was therefore taken by surprise when the faculty recommended him for an astronomy/mathematics teaching post in distant Austria. In a biographical account some years later, Kepler wrote, 'Nothing indicated that I had a talent for astronomy'.[2] In fact, his worst grade at Tübingen was in astronomy, an A–. But the faculty had evaluated him differently. The University Senate reported that he had 'such an outstanding and magnificent mind that something special may be expected of him'.[3] They must have known that he was already intrigued with the still-novel heliocentric system.

As a high school teacher at a provincial Lutheran school in Graz, Kepler's exploring mind stumbled on the idea that the Platonic solids could provide the spacing of the planets in the heliocentric system. With Maestlin's help, in 1596 he prepared a little book about his ideas, the *Mysterium cosmographicum* or Sacred Mystery of the Cosmos. With encouragement from the University Senate, Kepler included a copy of the *Narratio prima* as an appendix to his book. This was the preliminary tract from 1540 by Copernicus' only student, Georg

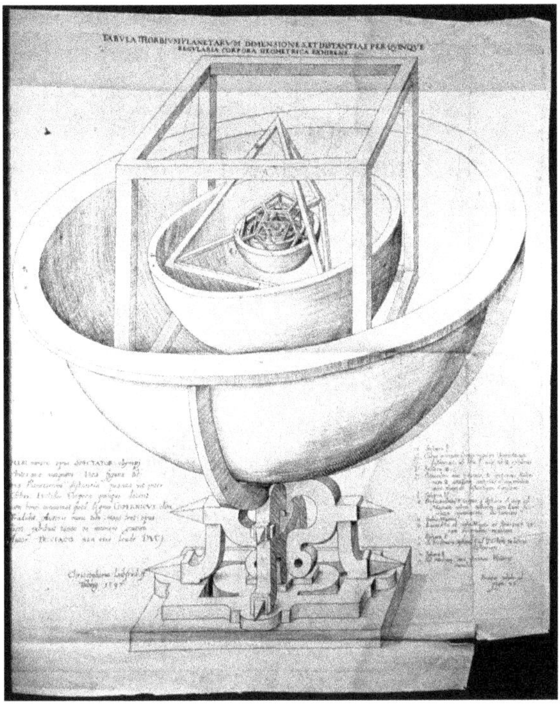

Figure 4.1 The five Platonic solids inserted between the planetary spheres as illustrated in Kepler's *Mysterium Cosmographicum*, 1596.

Joachim Rheticus. Kepler's *Mysterium* was the first enthusiastic heliocentric book since *De revolutionibus* itself. To his teacher, Kepler wrote: 'I wanted to become a theologian. For a long time I was restless. Now, however, behold how through my effort God is being celebrated in astronomy.'[4]

Kepler sent a couple copies of his book with a friend travelling to Italy, with instructions to give them to anyone who might be interested. Thus a copy came into the hands of Galileo, who wrote back to Kepler saying that he was a Copernican, though secretly. Kepler immediately responded, urging Galileo to stand forth with his views, but for over a decade Galileo remained a covert heliocentrist. Then, with Galileo's learning of the telescope, his converting a carnival toy into scientific discovery machine, and the publication in 1610 of his *Sidereus nuncius*, they got in touch again, but just barely. These two men became the most significant persons in the span between Copernicus and Newton for persuading the educated public to accept the heliocentric view as physical reality. But for sectarian reasons the Catholic Galileo and the Lutheran Kepler remained at arms length.

Meanwhile, in 1577 the Counter Reformation had become powerful in Graz, and Kepler along with the other Protestant teachers were given until sundown to leave town. Happily a copy of his *Mysterium* had reached Tycho Brahe, who eventually left Denmark for Prague. Impressed by the little book, Tycho offered Kepler a position as an apprentice in his enterprise. Tycho's move to Prague was, in Kepler's theological understanding, Divine Providence, because Denmark would have been too far away to move with his young family.

It was also Divine Providence in Kepler's longer view because the great observer Tycho had brought with him to Prague the most extensive and accurate set of planetary positions ever made. It was these observations, to which we shall return, that became grist for Kepler's mill in establishing a prediction procedure two orders of magnitude better than had been hitherto accomplished. But for now let us skip ahead to Kepler's longest book, his *Epitome of Copernican Astronomy*, published in three parts in 1618–21.

The *Epitome* contained a mature presentation of his life's work and the theoretical underpinning for his last great masterpiece, the *Rudolfine Tables* (Figure 4.2). The middle section, Book IV, is subtitled 'Celestial Physics.' Written as a catechism, he asks, 'What do you judge to be the principal parts of the world?' to which he replies:

> The philosophy of Copernicus counts up the principal parts of the world by dividing the form of the universe into parts. For in the sphere, which was the image of God the Creator and the Archetype of the world, there are three parts, symbols of the three persons of the Holy Trinity—the center, of the Father; the surface, of the Son; and the intermediate space, of the Holy Spirit.[5]

To most modern readers this answer looks like Kepler at his nuttiest. Actually it is an extremely pivotal concept for the Trinitarian Kepler, exactly what he is all about and central to our understanding of his motivations. We must remember that Kepler was trained as a theologian, not as an astronomer.

The very next section in Kepler's catechism contains an objection, which gives another clue to his reasoning. He proposes a rhetorical rebuttal to his first question, 'I thought the principal parts of the universe are reckoned to be the heavens and the earth,' and he proceeds to dismantle the rebuttal. The complaint itself summarized the traditional view of the universe. The dichotomy between the celestial and the terrestrial, with their differing elemental structure and physical laws, was a hinge point in Aristotle's logically constructed cosmology, but was what in retrospect we can call Aristotle's error. The traditional two-fold universe was already under erosion by Copernicus' theory and Tycho Brahe's

Figure 4.2 Kepler, as shown on the frontispiece of the *Rudolphine Tables*, 1627.

observations. In the *Epitome* Kepler's memorable warfare on Mars (the *Astronomia nova*) was being placed in the broader context of the revolutionary New Astronomy. In his catechetical response, he attacked the Aristotelian dichotomy, but he reminded his readers that: 'I am speaking now of the Earth insofar as it is part of the edifice of the universe, and not of the dignity of the governing creatures that inhabit it.'[6]

Despite his dissatisfaction with the Aristotelian dichotomy, Kepler as theologian must have found at least one part of Aristotle's cosmology particularly congenial, where the Stagirite says the love of God made the heavens go around.[7] In his *Metaphysics* Aristotle states that God as goodness expresses his love by giving motion to the cosmos, whirling the heavens so that the starry firmament spins on its axis every 24 hours. Closer to the fixed earth the planets are retarded, so that the nearby moon goes slowest; it appears to us to move most swiftly, but that is because we are envisioning the lunar motion with respect to the stars, not the earth's geocentric frame.

How can Kepler the theologian transfer the love of God to the heliocentric system? Copernicus has written: 'We see the sun as though on a royal throne governing the planets in his retinue as they circle round him.'[8] This harmonious picture must have provided a principal motivation for Copernicus as soon as he discovered that in the sun-centred arrangement Mercury, the fastest planet, automatically orbits closest to the sun, while lethargic Saturn provided the distant ancient border for the planets. It was an elegant aesthetic blue print; literally Copernicus had invented the solar system.

So if you are Kepler, the Copernican astro-physicist, you must ask, 'Where does the power come from to keep the planets moving?' Since in the Copernican system the outermost starry frame is fixed and the closer the planets are to the sun, the faster their periods of revolution are, the source of motion should be at the centre. Thus the sun must be causing the planets to wheel about in the sky. Hence it seems a perfectly natural thing for Kepler to adapt the Aristotelian concept (the source of motion coming from God), to identify metaphorically the sun at the middle as God bringing the action in the Copernican system. This Trinitarian image of the solar system must have been one of the strongest reasons for Kepler's acceptance of a sun-centred cosmos.

Kepler had become fascinated early on with this metaphorical embodiment of the Trinity in the heliocentric cosmology, but this was only a starting point. As it happened, Kepler got a second-hand copy of Copernicus' *De revolutionibus*, and the previous owner had written in it a sparse but extremely interesting set of small marginal notes. One of the longest of these came from Rheticus, Copernicus' only disciple and the man who persuaded his teacher to publish the book, though Kepler probably did not know its source. The note concerned the centre of the universe: whether it was within the sun itself, or whether it the centre of the earth's eccentric orbit, the point that Copernicus himself used for convenience. Rheticus' marginal note says he has discussed this in his *Narratio prima*, but 'my teacher skipped over it'. Now in Kepler's copy there is another note directly below that one, but in the hand of his teacher Michael Maestlin. Consequently we know that Kepler brought his book to his teacher, and they discussed this passage, concerning which point was the centre of the universe. Evidently they agreed on the importance of this topic, so when Kepler went to work with Tycho Brahe (who had assigned him to work on the problems with Mars), one of the first things he did was to ask permission to use as the fundamental reference point the sun itself rather than the centre of the earth's orbit. The centrality of the sun as the metaphorical symbol of God the Father was part of Kepler's theological stance, driving him toward a physical realism as he took up the problem of Mars.

Kepler's working notebook from his earliest days with Tycho Brahe still survives, and there we can see on the very first page the calculations to shift the reference point from the centre of the earth's orbit to the sun.[9] But the whole page is crossed off! In retrospect it is very amusing, because Kepler had employed extremely clumsy mathematics to try to work this out, and we can imagine that Tycho's chief assistant, Longomontanus, was looking over Kepler's shoulder and said: 'Young man, we have a manual of trigonometry here that will show you a much easier way to do that.'

We see on the second page that Kepler solved the problem with an algorithm straight out of Tycho's private manual of triangles. So this was the starting point, to get the sun physically into the centre. This step was very important, but Kepler immediately saw something else that was inconsistent. If the sun is the source of movement of the planets, then the earth ought to go faster when it is closer to the sun and slower when the earth is at aphelion. But as Copernicus had organized it, the earth's speed was always constant. Now it was known from antiquity that the seasons were not equal in length. In the northern hemisphere, summer was a few days longer than winter. This could be accounted for by assuming the sun moves faster in winter, but in Ptolemy's model the sun moved at a constant speed, so he adjusted the length of the seasons by placing his theoretical sun in an eccentric orbit. Copernicus, like Ptolemy, took the unequal length of seasons into account by using an eccentric orbit for the earth. Kepler realized that if he reduced the eccentricity of the earth's orbit, then a non-constant speed for the earth could partially account for the inequality of the seasons. The question then was, what proportion of the inequality of the seasons was accounted for by a non-constant speed of the earth, and what by the eccentricity of the orbit? This could be determined observationally if Kepler could measure the changing distance of the earth from the sun. He did try, for example, to establish the changing apparent diameter of the sun (which is inversely proportional to the distance), but unfortunately he was unable to measure the diameter with sufficient precision. (Later in the century G.D. Cassini used the solar projection in the Bologna Cathedral to verify the effect.[10])

Later astronomers promptly recognized that Kepler's account of his researches in the *Astronomia nova* was truly unusual because for the first time an astronomer explicitly recorded his struggle in dealing with a multiplicity of error-bound observations – a method of 'votes and ballots'. But for many generations, scholars assumed that the book was a full, pleading autobiographical account including all the pathways and blind alleys of Kepler's warfare on Mars. In fact, now that an examination of his manuscript archive has been realized, it is possible to see that the *Astronomia nova* is a highly organized rhetorical presentation.[11] As an example, there is no extended presentation of deciding what proportion of the seasonal inequality is to be assigned to the eccentricity of the earth's orbit and what proportion to the varying speed of the earth itself. Instead, the account goes first to a remarkably accurate solution of Mars' orbit where the proportion is a variable, and then he exclaims, 'Who would have believed it! The hypothesis, so closely in agreement with the observations, is nonetheless false.'[12]

Kepler's hypothesis agreed spectacularly well with the observations of longitude, within 2' of arc, essentially the limit of Tycho Brahe's naked-eye observations. Nevertheless, it disagreed with Tycho's observed latitudes, the angular position north or south of the ecliptic plane, related to the tilt of Mars's orbit. As a physical realist, Kepler was unprepared to use two different models, one for the longitudes and another for the latitudes (as Ptolemy and Copernicus had both done). But when he revised the model to take care of the latitudes, the longitudes went out by as much as 8'. Earlier, Copernicus had told his solitary disciple, Joachim Rheticus, that he would be very happy if his system could predict positions to 10'; he never knew that his tables for Mars occasionally had errors exceeding 250' (more than 4°).[13]

So, it was back to the drawing board for Kepler. He wrote: 'The divine benevolence has vouchsafed us Tycho Brahe, a most diligent observer, from whose observations the 8' error is shown, it is fitting that we should with thankful mind both acknowledge and honor this benefit of God . . . Because they could not be ignored, these eight minutes alone have led the way to the entire reformation of astronomy.'[14]

This famous quotation has engendered a certain amount of myth-making about the historical path of Kepler's achievements, not to mention the role of their Trinitarian underpinnings. When he became convinced about what is technically called 'the bisection of the eccentricity', which was necessary to predict the correct latitudes of Mars and which in the case of the earth's orbit meant that the inequality of the seasons is half-and-half from the eccentricity of the orbit and the variable speed, then he knew he could use Tycho Brahe's carefully determined eccentricity for the solar orbit simply by cutting it in two. Thus he did not have to find a delicate method to determine the sun's distances, though he needed at least some observational evidence to convince his readers that some readjustment was in fact necessary. He could do it roughly, but good enough, by a clever but relatively inaccurate system of triangulation. When he finally corrected the position of the earth's orbit, the occasional large errors in the prediction of Mars' positions dropped from over four degrees to half a degree, a huge achievement that is always outshone by his finding Mars's elliptical orbit, which reduced the errors by a further order of magnitude.

Part of the myth-making arises from a rather famous diagram in the *Astronomia nova* that shows the triangulation to Mars from various observing positions on the earth's orbit (Figure 4.3). At first glance this would seem to be an excellent way to locate a series of Martian positions through which a curve could then be drawn. Here there are two problems. First, related to yet another myth, Tycho did not observe the planetary positions routinely. He was very much directed toward specific observational goals, so he undertook observation campaigns every time Mars made a close approach, that is, about every two years, but in between he measured Mars rather infrequently. Thus Kepler could find only a handful of triangulation sets. Second, the triangulation method looks wonderful in the diagram, but in reality it is not in itself accurate enough to distinguish between a circle and an ellipse. (The difference between an eccentric circle and the corresponding ellipse is less than the thickness of the orbital line in the

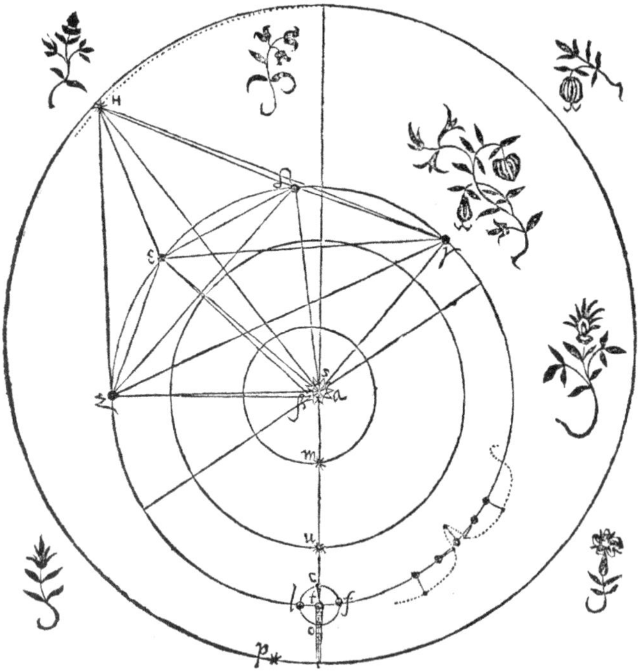

Figure 4.3 Kepler's Mars triangulation diagram from his *Astronomia nova*, 1609.

Mars returns to the place marked *H* every 687 days. In the time it takes Mars to return, the earth starting from point γ will revolve clockwise not quite twice, reaching point δ, and so on. In principle the method could be used to find the deviation of Mars' path from a circle, but as explained in our text, the deviation is too subtle to be gauged in this way. Nevertheless, it can be used to find the position of the earth's eccentric orbit.

woodblock diagram, contrary to what the diagram appears to show.) Nevertheless, the triangulation procedure, while inadequate to demonstrate the shape of Mars' orbit, was good enough to show that the earth's orbit needed to be repositioned with only half of its formerly assumed eccentricity.

The further steps to the elliptical form of the orbit are far more difficult to describe, in part due to the fact that Kepler examined multiple possible ovals that actually would have fit the data equally well. However, Kepler the realist was committed to finding a curve that he could explain in terms of physical causes. At the basis of his reasoning was the fact that he had already established as fundamental to his Trinitarian cosmology, namely, that the closer a planet is to the source of motion (the sun, the metaphorical symbol of God the Father), the faster it would go. Anachronistically, we can call this his distance law. It was difficult for him to handle this analytically with an eccentric circle, so he devised what we call his law of areas as a substitute. By dividing the area of the circle into many thin triangles he soon discovered that he accumulated too much area, and the eccentric circle had to be flattened slightly on the sides. That is to say, it was difficult observationally to pinpoint the distance of dots on the orbital path, but the integration over time related to angles, which were more readily observed. When Kepler noticed that an approximating ellipse had the sun at one focus, he exclaimed that it was like awakening from a deep sleep, and he knew he had a justifiable curve even though his physics (which lacked the concept of inertia) was simply wrong. It was this inspired guess-work, motivated by his Trinitarian framework, that won the day. His 'aha! moment' is captured by his word *focus*, literally 'hearth', for the sun was the hearth of his universe.

In the *Astronomia nova* Kepler gave what are now called the first and second laws of planetary motion (the ellipse and the law of areas). He did not point them out as laws, and never spoke of 'laws of nature'. These two plus the so-called harmonic law that Kepler published in 1619 were singled out and numbered for the first time in 1774 by the French astronomer J.J. Lalande.[15]

Where then does the concept of laws of nature come from? In the modern sense, they stem from Descartes. He wanted to build a philosophy from scratch, 'I think therefore I am' and so on. Repelled by the 'superstitious notion' of action at a distance, Descartes formulated a plenum universe filled with a mechanical aether, and then got stuck wondering where all the motion came from that made his plenum universe function. He concluded that ultimately it must come from divine creation. Thus, as John Henry and Peter Harrison have shown, the modern sense of 'law of nature' is intimately connected with divine law.[16] Boyle and

Newton also spoke of laws of nature, which for them were likewise embedded in a divine context.

Isaac Newton, who was equally baffled about action at a distance, says 'concerning gravity itself, I feign no hypotheses' and that space is the 'sensorium of God' related to the ultimate nature of the universe.[17] It is like Kepler identifying the intermediate space with the Holy Spirit, or Descartes having God fill the universe with a plenum in motion at the beginning. This search for the fundamental stuff of the cosmos may still be our quest, ultimately filling the universe, perhaps with dark energy!

Notes and references

1. The principal part of this essay was originally published in *Science and Theology*, **9**(1), 2011, 45–51; we thank Taylor and Francis for permission to reprint that part.
2. *KGW* (= *Johannes Kepler Gesammelte Werke*), Vol. 3, 108.
3. *KGW*, Vol. 13, 4.
4. *KGW*, Vol. 13, 40.
5. Translation modified from *Great Books of the Western World*, Vol. 16, Encyclopaedia Britannica, 1952, Johannes Kepler, *Epitome of Copernican Astronomy*, 1618–21, pp. 853–4. (Note that in the current second edition this work is in Vol. 15.)
6. Kepler, *Epitome*, p. 854.
7. Aristotle, *Metaphysics*, Book XII, chapter 7.
8. Nicolaus Copernicus, *De revolutionibus*, Nuremberg, 1543, Book 1, chapter 10.
9. See James Voelkel, *The Composition of Kepler's Astronomia nova*, Princeton University Press, 2001, pp. 100–1. The page in Kepler's workbook is partially transcribed in *KGW*, Vol. 20:2 (1998), 18, but the drama of the crossed-out part is entirely missing. The page is reproduced in Owen Gingerich, *The Eye of Heaven*, American Institute of Physics, 1993, p. 335.
10. John Heilbron, *The Sun in the Church*, Harvard University Press, 1999, pp. 107–8 and 296.
11. In 1970 I finally received a microfilm of the relevant papers in the Kepler archive at the Russian Academy of Sciences Archive in Leningrad, which I have shared with D.T. Whiteside, William Donahue, and James Voelkel. See my Kepler versus the computer revisited, in F. Krafft, K. Meyer, and B. Sticker (eds), *Proceedings of the Internationales Kepler-Symposium, Weil der Stadt 1971*, Gerstenberg, 1973, pp. 307–14.
12. Johannes Kepler, *Astronomia nova*, Prague, 1609, beginning of chapter 19; translation modified from William Donahue (trans.), *Johannes Kepler: New Astronomy*, Cambridge University Press, 1992, p. 281.

13. Owen Gingerich, The great Martian catastrophe and how Kepler fixed it, *Physics Today*, **64**(9), 2011, 50–4.

14. Kepler, *Astronomia nova*, end of chapter 19; Donahue, *New Astronomy*, p. 286.

15. J.-J. LaLande, *Abrégé d'Astronomie*, Paris, 1774, paragraph 467, 201.

16. John Henry, Metaphysics and the origins of modern science: Descartes and the importance of laws of nature, *Early Science and Medicine*, **9**(2), 2004, 73–114. Peter Harrison, The development of the concept of laws of nature, chapter 2 in Fraser Watts (ed.), *Creation: Law and Probability*, Fortress with Ashgate, 2008, pp. 13–36.

17. Isaac Newton, *Philosophiae naturalis principia mathematica*, Cambridge, 1713, Scholium Generale, 484, 'Hypotheses non fingo'. Newton's reference to 'sensorium of God' appears in Query 31 of his *Opticks*, London, 1704, and was brought out as significant by Joseph Addison who referenced it in *Spectator*, **565**, July 1714 (which is cited by the *Oxford English Dictionary*).

CHAPTER 5

The Lull before the storm: combinatorics in the Renaissance

ROBIN WILSON AND JOHN FAUVEL

Early combinatorics

Combinatorial mathematics can be traced back through some 3000 years and arose from a number of concerns, both theoretical and practical.[1] In this chapter we show how combinatorial ideas first arose and were explored in the West. As we shall see, religious ideas were a strong motivating component of early combinatorial explorations.

The earliest examples that we can identify as combinatorial arose in China and India, while the first European works were in the traditions of Jewish and Christian religious and theological concerns. We set the scene by outlining the earliest contributions to the subject, and then describe the seminal developments in early modern Europe – notably, the writings of the thirteenth-century Catalan mystic Ramon Lull (or Llull), which greatly influenced combinatorics as it took shape throughout the Renaissance. In particular, Lullism was a major inspiration for the combinatorial work of two seventeenth-century priests, Marin Mersenne and Athanasius Kircher, and for that of Leibniz in his early years.

As we mentioned, combinatorial mathematics originated in the ancient civilizations of China and India.[2,3] According to legend, Emperor Yu was standing on the bank of the River Luo, a tributary of the Yellow River, when a sacred turtle emerged from the river with the numbers 1 to 9 on its back in a square array: this arrangement, called the *Luoshu*, had the 'magic' property that the sum of the numbers in any row, column, or diagonal is the same, 15 (see Figure 5.1a). Over the centuries such magic squares came to acquire religious and mystic significance and many examples were constructed. In India, a statue of the Indian god Vishnu shows him holding in his four hands a discus, conch, lotus, and mace (see Figure 5.1b): since his first hand can hold any of the four objects, his second hand can hold any of the remaining three, and so on, there are $4! = 24$ possible permutations. A similar ancient problem concerns the ten-handed god Sambhu, referred to by Bhaskara II:[4]

> How many are the variations of form of this god by the exchange of his ten attributes held in his ten hands: the rope, the elephant's hook, the serpent, the tabor, the skull, the trident, the bedstead, the dagger, the arrow and the bow?

The answer is 10!: over three million ways!

Throughout the millennia many writers in many places have been interested in the combinatorics of poetic metre. Such concerns certainly featured in ancient India, and were frequently religious in content. Even as late as 1617 the Flemish professor Erycius Puteanus considered a famous eight-word Latin hexameter line by the Jesuit priest Bernard Bauhuis in praise of the Virgin Mary:[5]

> Tot tibi sunt dotes, Virgo, quot sidera caelo.

> (Thou hast as many virtues, O Virgin, as there are stars in heaven.)

In total, there are $8! = 40{,}320$ permutations of these eight words, but each permutation was required to fit the basic hexameter pattern:

> dum-diddy, dum-dum, dum-dum, dum-dum, dum-diddy, dum-dum;

for example:

> Tot dotes tibi, quot caelo sunt sidera, Virgo.

> Sidera quot caelo, tot sunt Virgo tibi dotes.

> Quot caelo sunt sidera, tot Virgo tibi dotes.

> Sunt caelo tot Virgo tibi, quot sidera, dotes.

Since Mary clearly had many more virtues than the number of stars in the universe, Puteanus then proceeded to write out 1022 (the number in Ptolemy's star catalogue) of these permutations in full.

(a)

(b)

Figure 5.1 (a) The Luoshu magic square and (b) the Indian God Vishnu. ((a) From Cheng Dawei, *Suanfa tongzong* (1592) (reprinted by Taishantung, Jingdu, 1882). (b) From S.G. Goodrich, *Lights and Shadows of Asiatic History*, Bradbury, Soden & Co., Boston, 1844.)

Another fertile source of early combinatorial work lies in the Jewish tradition.[6] An important sacred text from before the eighth century was the mystical *Sefer Yetsirah* (Book of Creation), in which are calculated the numbers of ways that the 22 letters of the Hebrew alphabet can be arranged, subject to various rules. Jewish mystics believed that God created the world and everything in it by naming them all, using only the 22 letters of the Hebrew alphabet:

> God drew them, combined them, weighed them, interchanged them, and through them produced the whole creation and everything that is destined to be created.

In particular, the unnamed author discovered the number of words with only two letters:

> [God] combined the aleph [the first letter] with all the other letters in succession, and all the other letters again with aleph; bet [the second letter] with all and all again with bet, and so the whole series of letters.

The total number of pairs is $22 \times 21 = 462$, and since the ordering is immaterial we take half of this, giving the answer 231. The author then went on to count permutations of letters (see Figure 5.2), continuing up to 11!, since 11 was the greatest number of different letters in any word appearing in the Hebrew Bible.

A later (and more detailed) study of permutations and combinations is found in the work of the Spanish–Jewish scholar Rabbi Abraham ibn Ezra (*c.* 1090–*c.* 1167), who calculated the number of possible conjunctions of the planets, in an astrological context[7]; in effect, he was exploring a general rule for the binomial coefficients $C(n, k)$, where

$$C(n, k) = \frac{n \times (n-1) \times (n-2) \times \ldots \times (n-k+1)}{1 \times 2 \times \ldots \times k}.$$

דברים, דבירם, דבימר, דבמיר, דבמרי, דברמי, דרבים, דרבמי, דרימב, דריבם, דרמבי, דרמיב,
דיברם, דירבם, דירמב, דיבמר, דימרב, דימבר, דמביר, דמברי, דמרבי, דמריב, דמיבר, דמירב,
בדימר, בדירם, בדרמי, בדרים, בדמרי, בדמיר, בירמד, בידרם, בירדם, בימדר, בימרד, בידמר,
במדרי, במדיר, במריד, במרדי, במדיר, במירד, בדמיר, בדמרי, בדרמי, בדרים, בדימר, בדירם,
רימדב, רימבד, רידבם, רידמב, ריבדם, ריבמד, רדמבי, רדבמי, רדיבם, רדימב, רדבים, רדמיב,
רבמדי, רבדמי, רבדים, רבידם, רבימד, רבמיד, רמידב, רמיבד, רמדיב, רמביד, רמבדי, רמדבי,
ימדרב, ימרדב, ימבדר, ימברד, ימרבד, ימדבר, ידברם, ידרבם, ידבמר, ידמבר, ידרמב, ידמרב,
יבדרם, יבדמר, יברדם, יברמד, יבמדר, יבמרד, ירדבם, ירדמב, ירמבד, ירבמד, ירבדם, ירדמב,
מדברי, מדרבי, מדביר, מדיבר, מדירב, מדריב, מבדיר, מבידר, מבריד, מבירד, מברדי, מבדרי,
מרדבי, מרבדי, מרדיב, מרביד, מריבד, מרידב, מידבר, מידרב, מיברד, מיבדר, מירדב, מירבד

Figure 5.2 The 5! = 120 permutations of five Hebrew symbols, from a medieval commentary on the Sefer Yetsirah. (S. Donnolo, Hakhmoni (*Commentary on Sefer Yetzirah*), p. 946 (reprinted in Warsaw, 1884).)

In particular, he announced that 'the sum of the numbers from 1 to any desired number is found by multiplying it by half of itself and by half of unity' – that is,

$$1+2+\ldots+n=n\times\left(\tfrac{1}{2}n+\tfrac{1}{2}\right)=\tfrac{1}{2}n(n+1)$$

in modern algebraic notation. Among his astrological observations were calculations involving conjunctions of the seven planets (including the sun and the moon), because these were believed to have a profound influence on human life:

> [For conjunctions of two planets] It is known that there are seven planets. Now Jupiter has six conjunctions with the [other] planets. Let us multiply then 6 by its half and by half of unity. The result is 21, and this is the number of binary conjunctions [that is, $C(7, 2) = 21$].

> [For conjunctions of three planets] We begin by putting Saturn with Jupiter and with them one of the others. The number of the others is five: multiply 5 by its half and by half of unity. The result is 15. And these are the conjunctions of Jupiter.

He then found the number of ternary conjunctions involving Saturn but not Jupiter (which is $C(5, 2) = 10$), and then Mars but neither Jupiter nor Saturn, and concluded with the sum:

$$C(7,3)=C(6,2)+C(5,2)+C(4,2)+C(3,2)+C(2,2)$$
$$=15+10+6+3+1=35.$$

He also used a similar approach to calculate the number of quaternary conjunctions, again obtaining the correct answer, $C(7, 4) = 35$.

Ramon Lull (c. 1232–1316)

The most important combinatorial influence from around this time – and for several centuries thereafter – was also working within this richly eclectic Western Mediterranean culture. Ramon Lull was a scholar living on Majorca, who wrote extensively in Arabic, Latin, and his native Catalan. The main purpose of his writings was to develop materials for the conversion of Moslem infidels.

Lull aimed to unify all knowledge into a single system, and through this to teach Christian theology so logically that Moslems could not but see its truth and be converted. Regarding all reality as the embodiment of aspects of the divinity, he attempted in his *Ars Magna* (*c.* 1273) to teach the natural sciences of his day as congruent expressions of the truths of theology and philosophy. He believed that he had discovered 'an Art of thinking which was infallible in all spheres because based on the actual structure of reality, a logic which followed the true patterns of the universe'.[8]

Lull's method was based on the belief that knowledge arises from a finite number of basic principles or categories. By moving through all possible combinations of these categories, we reach all knowledge: *combinatorics is thus the basic tool for exploring all that can be known*. In particular, he used 'combinatory diagrams' to present the active manifestations of the divine attributes, which he called 'Dignities': these included *Bonitas* (goodness), *Potestas* (power), *Sapientia* (wisdom), and so on. One of his diagrams (Figure 5.3a) shows the relationships among nine of these Dignities, while another (Figure 5.3b) consisted of three revolving wheels that could be rotated independently so as to illustrate all the different combinations of three Dignities.

Lull used letters to stand for divine attributes and other things – sometimes complex concepts, phrases, or even sentences – which he then combined in a logical or combinatorial algebra, while the revolving wheels provide a mechanical solution, at least on a conceptual level, to the combinatorial problem of exploring all the possibilities. With these algebraic symbols, with the wheels, and with further devices such as colours, geometrical shapes, and binary oppositions, Lull was able to weave a combinatorial theological tapestry of fantastic complexity.

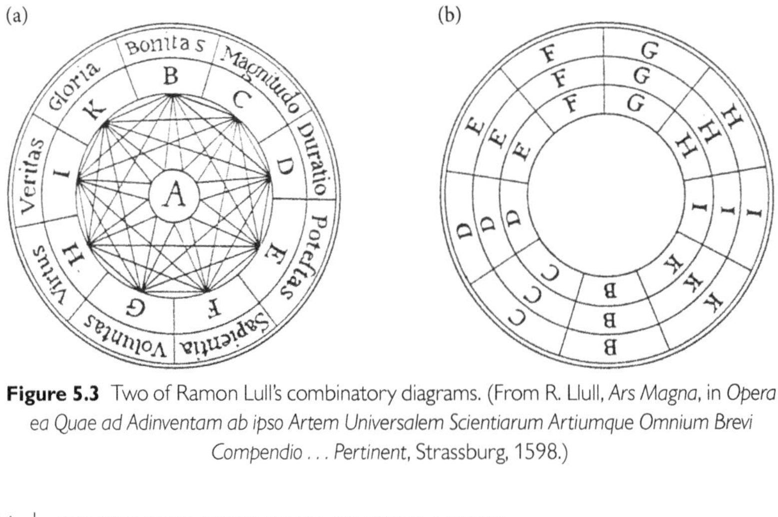

Figure 5.3 Two of Ramon Lull's combinatory diagrams. (From R. Llull, *Ars Magna*, in *Opera ea Quae ad Adinventam ab ipso Artem Universalem Scientiarum Artiumque Omnium Brevi Compendio . . . Pertinent*, Strassburg, 1598.)

Lull's influence

In the centuries following his death, Lull's influence spread throughout Europe – not universally, but in particular milieux. Although his ideas appealed only to a small minority, it was an influential one, and by the late Renaissance these ideas were of particular interest to all those engaged in unifying knowledge into a single system.

This activity was frequently associated with Christian proselytizing. Thus, to a surprising and unexpected degree, the many scholars of the sixteenth and seventeenth centuries who explored combinatorial ideas were religious of one persuasion or another – and frequently Jesuits. Among these we particularly note the work of

- the German Jesuit Christoph Clavius (1537–1612),
- the Swiss-born Jesuit Paul Guldin (1577–1643),
- the French Minimite Marin Mersenne (1588–1648),
- the Spanish Jesuit Sebastian Izquierdo (1601–81),
- the German-born Jesuit Athanasius Kircher (1601–80),
- the Spanish Cistercian Juan Caramuel (1606–82),
- the German Jesuit Gaspar Schott (1608–66),
- the Belgian Jesuit Andreas Tacquet (1612–60).

From the viewpoint of the history of mathematics, the number of clerics who discussed combinatorial ideas is surprising, as not every mathematician of the period had close religious links. On the other hand, the predominance of Jesuits can be explained by the high educational direction of that order, and also by the tendency of Jesuit teachers to write encyclopedic texts. By analysing the contexts in which combinatorial ideas were discussed, we can judge the attraction of this approach for religious teachers. Here we outline the ideas of two of the most interesting of these writers, Marin Mersenne and Athanasius Kircher; further information about the others can be found in work by the historian of science Eberhard Knobloch.[9,10]

Marin Mersenne (1588–1648)

Like his friend and near-contemporary René Descartes (1596–1650), Mersenne was educated at the new Jesuit College at La Flèche. He spent almost all his life at the Minimite convent in Paris, writing books and conducting research into a wide range of mathematical and scientific pursuits, such as number theory (the 'Mersenne primes') and the theory of sound. In those days before learned journals, he carried out a vast scientific correspondence with the leading scholars of Europe, taking a major clearing-house role in passing on new ideas and discoveries in mathematics and other branches of science. His efforts were devoted to the ultimate end of defending Christianity against heretics and unbelievers; even where he was discussing what to us are simply mathematical ideas, the whole enterprise had a religious location and ultimate grounding: he saw mathematics as a precise science that is invaluable for understanding theology and the Holy Scriptures. As Knobloch has remarked:[11]

> His religious interests were inseparably connected with his interest in the Llullistic combinatorial art, believing that God was the first to practise this fundamental universal art when he created the world, and that God was the first combinatorialist when he combined the single parts of the universe.

Mersenne's combinatorial ideas are found in six of his publications. Two of these refer in the same breath to 'mathematicians and theologians' as among the intended audience: these were his large encyclopedia *Quaestiones Celeberrimae in Genesim* (Most Famous Questions Relating to Genesis), whose lengthy title describes it as 'A Work Useful for Theologians, Philosophers, Physicians, Legal Advisors, Mathematicians, Musicians, But Especially for Those who are Dealing with Optical Reflections', and *La Vérité des Sciences Contre les Sceptiques ou Pyrrhoniens* (The Truth of the Sciences Against the Sceptics or Pyrrhonians), whose purpose was to show how useful mathematics is for the understanding of Holy Scripture. In these works, Mersenne explained how to calculate the numbers of permutations and combinations of a given number of objects, listed all of the 120 possible 'songs' with five notes (ut, re, mi, fa, sol), and presented a table containing every factorial number up to 50!.

Mersenne also wrote two major treatises on music, his *Harmonie Universelle* and *Harmonicorum Libri*, which contain several extended discussions of the various possible arrangements of consonances. In particular, he listed the 720 'songs' with six notes (ut, re, mi, fa, sol, la), writing them out fully in musical notation, and then presented, in a musical setting, the most extensive tables produced up to that time of numbers of permutations and combinations with and without repetition (see Figure 5.4), calculating, for example, the number of ways of selecting up to 12 musical notes from 36 – this number is over 12 billion. In these works he also listed all 60 arrangements of the letters in the name *IESUS*, and extended his table of factorial numbers up to 64!, a 90-digit number

Tabella pulcherrima & vtilißima Combinationis duodecim Cantilenarum.

I.	II.	III.	IV.	V.	VI.	VII.	VIII.	IX.	X.	XI.	XII.
1	1	1	1	1	1	1	1	1	1	1	1
2	3	4	5	6	7	8	9	10	11	12	13
3	6	10	15	21	28	36	45	55	66	78	91
4	10	20	35	56	84	120	165	220	286	364	455
5	15	35	70	126	210	330	495	715	1001	1365	1820
6	21	56	126	252	462	792	1287	2002	3003	4368	6188
7	28	84	210	462	924	1716	3003	5005	8008	12376	18564
8	36	120	330	792	1716	3432	6435	11440	19448	31824	50388
9	45	165	495	1287	3003	6435	12870	24310	43758	75582	125970
10	55	220	715	2002	5005	11440	24310	48620	92378	167960	293930
11	66	286	1001	3003	8008	19448	43758	92378	184756	352716	646646
12	78	364	1365	4368	12376	31824	75582	167960	352716	705432	1352078
13	91	455	1820	6188	18564	50388	125970	293930	646646	1352078	2704156
14	105	56[illegible]	2380	8568	27132	77520	203490	497420	1144066	2496144	5200300
15	120	680	3060	11628	38760	116280	319770	817190	1961256	4457400	9657700
16	136	816	3876	15504	54264	170544	490314	1307504	3268760	7726160	17383860
17	153	969	4845	20349	74613	245157	735471	2042975	5311735	13037895	30421755
18	171	1140	5985	26334	100947	346104	1081575	3124550	8436285	21474180	51895935
19	190	1330	7315	33649	134596	480700	1562275	4686825	13123110	34597290	86493225
20	210	1540	8855	42504	177100	657800	2220075	6906900	20030010	54627300	141120525
21	231	1771	10626	53130	230230	888030	3108105	10015005	30045015	84672315	225792840
22	253	2024	12650	[illegible]5780	296010	1184040	4292145	14307150	44352165	129024480	354817320
23	276	2300	14950	80730	376740	1560780	5852925	20160075	64512290	193536720	548354040
24	300	2600	17550	[illegible]8280	475020	2035800	7888725	28048800	92561040	286097760	834451800
25	325	2925	2047[illegible]	[illegible]875[illegible]	593775	2629575	10518300	38567100	131128140	417225900	1251677700

Figure 5.4 Marin Mersenne's arithmetical triangle. (From M. Mersenne, *Harmonicorum Libri XII*, Lutetiae Parisiorum (1636), p. 136.)

which was then the largest factorial number ever calculated, though several of his later values were incorrect.

Athanasius Kircher (1601–80)

Another 'one-man intellectual clearing-house', as he has been called, was the Jesuit priest Athanasius Kircher, who was born in Thuringia, Germany, and died in Rome. His many writings, greatly influenced by Lull, illustrate well the range of problems to which some scholars of the time were applying a combinatorial approach: for instance, in his *Polygraphia Nova et Universalis* (New and Universal Language) of 1663, he used the art of combinations to reduce all tongues to just one universal language.[12] This had a clear religious motivation: to repair the linguistic confusion, symbolizing human discord and mutual aggression, which followed the Tower of Babel. The project may seem inherently implausible to us, as it did to some linguists of the time (and to Leibniz also), but the exercise in combinatorics shows a growing confidence in this style of mathematatizing all the possibilities.

Kircher's range of interests was immense, and his books included extensive treatises on many topics ranging from Egyptian hieroglyphics and China to optics, acoustics, and magnetism. Here we concentrate on a single work – his vast *Ars Magna Sciendi Sive Combinatoria* of 1669, a 12-part system of logic derived from Ramon Lull. Its impressive frontispiece (see Figure 5.5) presents it as 'The Great Art of Knowledge, or the Combinatorial Art, Through Which the Broadest Door is Opened for Quickly Acquiring Knowledge in all Arts and Sciences . . .'. Shown here is the eye of God presiding over Theology, Metaphysics, Physics, Logic, Medicine, Mathematics, Moral Ethics, Ascetics, Jurisprudence, Politics, Scriptural Interpretation, Controversy, Moral Theology, Rhetoric, and the Combinatorial Art, while the Divine Sophia (the goddess of wisdom) holds a tablet of the 'Alphabet of the Art', featuring Lull's divine attributes. Underneath Sophia is a Greek inscription, saying 'Nothing is more beautiful than to know the All'.

Of the *Ars Magna Sciendi*'s 12 parts, the third, *Methodus Lulliana*, is a general description of Lullist principles. It is followed immediately by the 50-page Book IV, *Ars Combinatoria*, which commences with word arrangements, such as a list

Figure 5.5 The frontispiece of Athanasius Kircher's *Ars Magna Sciendi Sive Combinatoria*. Amsterdam (1669).

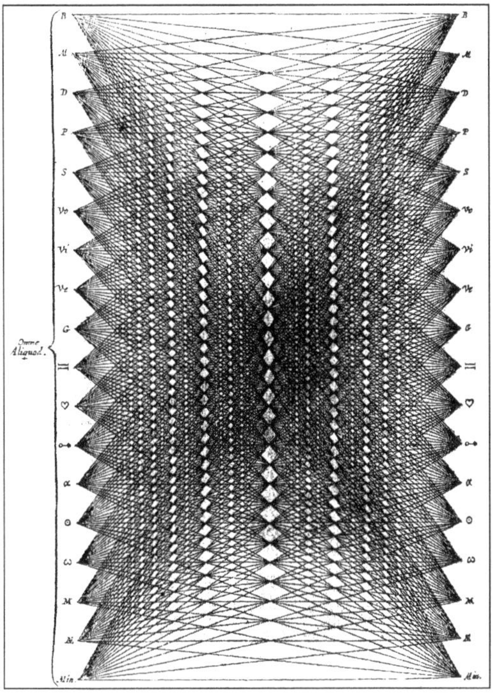

Figure 5.6 An 18 × 18 bipartite graph linking 18 of Lull's divine attributes in pairs. (From A. Kircher, *Ars Magna Sciendi Sive Combinatoria*, Johannes Janssonius a Waesberge, Amsterdam (1669), p. 170.)

of the 6 permutations of the letters in the word *ORA*, and the 24 permutations of the letters in the word *AMEN*. Kircher then observed that the word *PATER* yields 120 arrangements, and proceeded to give a table of factorials from 1! to 50!, copied from Mersenne. This is followed by an extended discussion of the ways in which one can select various of the Lullian dignities and attributes, and includes a table of numbers of ordered selections from up to ten objects. There is also a discussion of how to combine the 18 divine attributes in pairs and triples, including a square array showing all the 324 pairs of two attributes and a magnificent 18 × 18 bipartite graph (see Figure 5.6).

Gottfried Wilhelm Leibniz (1646–1716)

Leibniz produced his *Dissertatio de Arte Combinatoria* (Dissertation on the Combinatorial Art) in 1666, at the age of 20 (see Figure 5.7). In its scope and style it differs little from the earlier Lullist works of Mersenne, Kircher, and

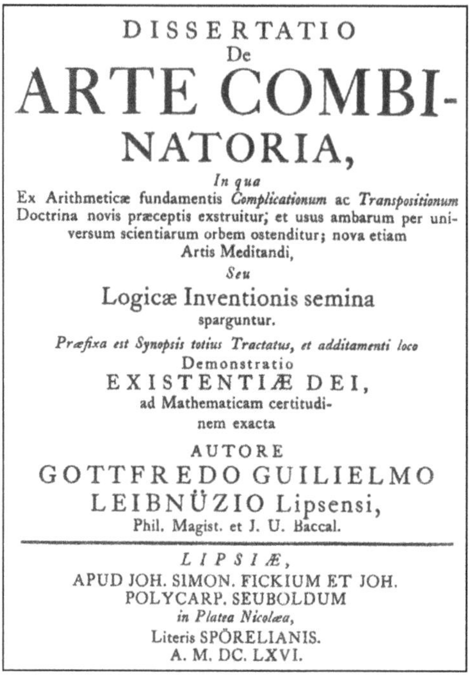

DISSERTATIO
De
ARTE COMBI-
NATORIA,
In qua
Ex Arithmeticæ fundamentis *Complicationum* ac *Transpositionum*
Doctrina novis præceptis exstruitur; et usus ambarum per universum scientiarum orbem ostenditur; nova etiam
Artis Meditandi,
Seu
Logicæ Inventionis semina
sparguntur.
Præfixa est Synopsis totius Tractatus, et additamenti loco
Demonstratio
EXISTENTIÆ DEI,
ad Mathematicam certitudinem exacta
AUTORE
GOTTFREDO GUILIELMO
LEIBNÜZIO Lipsensi,
Phil. Magist. et J. U. Baccal.

LIPSIÆ,
APUD JOH. SIMON. FICKIUM ET JOH.
POLYCARP. SEUBOLDUM
in Platea Nicolæa,
Literis SPÖRELIANIS.
A. M. DC. LXVI.

Figure 5.7 Gottfried Wilhelm Leibniz's *Dissertatio de Arte Combinatoria*. Fick & Seubold, Leipzig, 1666, title page = *Mathematische Schriften*, Vol. 5, 7.

others, being essentially a discussion of various problems involving permutations and combinations, and including as a bonus a proof of the existence of God 'with complete mathematical certainty'. The *Dissertatio* is discussed in detail by Knobloch.[13] In later years Leibniz recalled Lull's influence on this work:[14]

When I was young, I found pleasure in the Lullian art, yet I thought also that I found some defects in it, and I said something about these in a little schoolboyish essay called *On the Art of Combinations* . . . I have found something valuable, too, in the art of Lully and in the *Digestum Sapientiae* of the Capuchin, Father Ives, which pleased me greatly because he found a way to apply Lully's generalities to useful particular problems.

Once he had broken away from Lull's influence, Leibniz's combinatorial work made great strides, becoming more 'mathematical' and less 'philosophical', although he had a broader conception of combinatorics than we do today. In his unpublished *Nachlass* can be found work on partitions, determinants, symmetric functions, and probability theory – in particular, anticipating Stirling numbers of the second kind, Euler's recursion formula for the number of partitions of n into k parts, and some special results on partitions; these did not appear in print until 1840. Details of Leibniz's work in these areas can be found in Knobloch.[15]

After Leibniz, the Lullian influence was never again central to combinatorial explorations in mathematics. In a way, though, it moved into mathematics at an even deeper level, as one of the tributaries of the collection of influences that inspired Leibniz to work towards a calculus – that is, towards realizing the idea that there must be general truths of method that enable truth to be reached through a process of calculation and which underlie a host of disparate results. The Leibniz who discovered the calculus in the 1670s (later than, and in a different form from, Isaac Newton) was someone whose mind had been reinforced in this direction by reflecting upon the Art of Ramon Lull.

Conclusion

Combinatorics is for many the epitome of cool and detached analytical rational thought. It is interesting, therefore, to reflect that a significant part of its early development was in a context of religion's mystical dimensions. The intellectual tradition in which it flourished in Europe for its first four or five centuries was decidedly irrational, if not anti-rational, in overall tone, by comparison with the more rational, but less mathematical, traditions from the classical past.

There is a gentle irony in reflecting that the art which once inspired a more serious attention to combinatorics than any other intellectual current of the Middle Ages and Renaissance ended up being satirized in the pages of Jonathan Swift's great novel *Gulliver's Travels* (1726).[16] In Part III Gulliver encounters a professor in the Laputa Academy who had devised a large wooden frame divided into small squares with pieces of paper pasted on them (see Figure 5.8):

> . . . on these papers were written all the words of their language, in their several moods, tenses, and declensions; but without any order . . . The [forty] pupils, at his command, took each of them hold of an iron handle, whereof there were forty fixed around the edges of the frame; and giving them a sudden turn, the whole disposition of the words was entirely changed. He then commanded six-and-thirty of the lads, to read the several lines softly, as they appeared upon the frame; and where they found three or four words together that might form part of a sentence, they dictated to the four remaining boys,

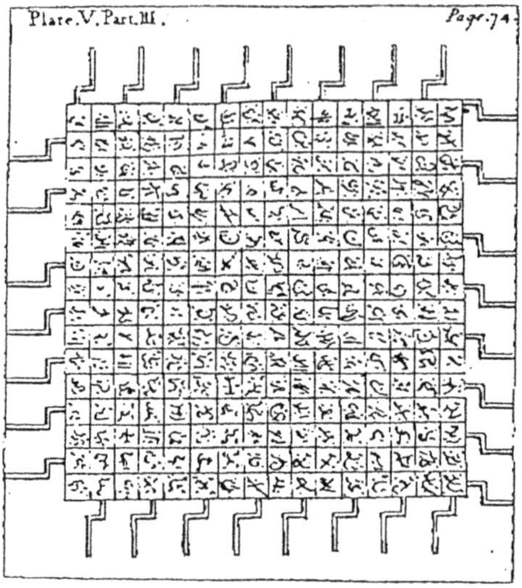

Figure 5.8 The square frame in Part III of *Gulliver's Travels*. (From J. Swift, *Travels into Several Remote Nations of the World. In Four Parts. By Lemuel Gulliver, First a Surgeon, and then a Captain of several Ships* [*Gulliver's Travels*], Part III, A Voyage to Laputa, Benjamin Motte, Ireland, 1726.)

who were scribes . . . Six hours a day the young students were employed in this labour, and the professor showed me several volumes in large folio already collected, of broken sentences, which he intended to piece together, and out of those rich materials, to give the world a complete body of all arts and sciences . . .

The explanation of its purpose, for displaying different combinations of words on successive turns of the handles, is all too reminiscent of the combinatorial tradition stemming from Lull's method: by this contrivance (emphasis added):

the most ignorant person, at a reasonable charge, and with a little bodily labour, might write books in philosophy, poetry, politics, laws, *mathematics, and theology*, without the least assistance from genius or study.

Acknowledgement

This chapter is an enlarged and updated version of the following article, written for a combinatorial readership: John Fauvel and Robin J. Wilson, The Lull before the storm: Combinatorics and religion in the Renaissance, *Bulletin of the Institute of Combinatorics and its Applications*, **11**, 1994, 49–58.

Notes and references

1. See Robin Wilson and John J. Watkins (eds), *Combinatorics: Ancient & Modern*, Oxford University Press, 2013.
2. See Takanori Kusuba and Kim Plofker, Indian combinatorics, in Wilson and Watkins (eds), *Combinatorics*, chapter 1, pp. 40–62, and Andrea Bréard, China, in Wilson and Watkins (eds), *Combinatorics*, chapter 2, pp. 64–81.
3. N.L. Biggs, The roots of combinatorics, *Historia Mathematica*, **6**, 1979, 109–36.
4. Bhaskara II, *Algebra, with Arithmetic and Mensuration, from the Sanscrit of Brahmegupta and Bhascara*, trans. H.-T. Colebrook, John Murray, 1817.
5. E. Puteanus, *Pietatis Thaumata*, Antwerp, 1617; see also Donald E. Knuth, Two thousand years of combinatorics, in Wilson and Watkins (eds), *Combinatorics*, Introduction, pp. 2–37.
6. See Victor J. Katz, Jewish combinatorics, in Wilson and Watkins (eds), *Combinatorics*, chapter 4, pp. 108–21.
7. See Jekuthiel Ginsburg, Rabbi ben Ezra on permutations and combinations, *The Mathematics Teacher*, **15**, 1922, 347–56.
8. See F.A. Yates, *Lull and Bruno*, Routledge and Kegan Paul, 1982, on p. 12.
9. Eberhard Knobloch, Musurgia universalis: unknown combinatorial studies in the age of Baroque absolutism, *History of Science*, **17**, 1979, 258–75.
10. See Eberhard Knobloch, Renaissance combinatorics, in Wilson and Watkins (eds), *Combinatorics*, chapter 5, pp. 122–45.
11. Knobloch, Renaissance combinatorics.
12. Joscelyn Godwin, *Athanasius Kircher: A Renaissance Man and the Quest for Lost Knowledge*, Thames and Hudson, 1979.
13. Eberhard Knobloch, The mathematical studies of G. W. Leibniz on combinatorics, *Historia Mathematica*, **1**, 1974, 409–30.
14. G.W. Leibniz, *Gottfried Wilhelm von Leibniz: Philosophical Papers and Letters*, Vol. 2, ed. L.E. Loemker, University of Chicago Press, 1956, on p. 1067.
15. See Eberhard Knobloch, The origins of modern combinatorics, in Wilson and Watkins (eds), *Combinatorics*, chapter 6, pp. 146–65.
16. J. Swift, *Gulliver's Travels*, Part III, A Voyage to Laputa, Penguin Books, 1967 (original edn 1726).

CHAPTER 6

Mystical arithmetic in the Renaissance: from biblical hermeneutics to a philosophical tool

JEAN-PIERRE BRACH

More to it than meets the eye: qualitative number

Before launching into an overview of the history of number symbolism in the Early Modern period (fifteenth–seventeenth centuries) we must first define the term 'mystical arithmetic' in the title of this chapter, and second dispose of a historical route between the early Christian church and antiquity which does not exist.

First to the definition: 'mystical arithmetic', or arithmology, means a type of understanding of mathematics that imbues numbers with the capacity of signifying more than just the quantity they materially refer to.[1] Interest in such a *qualitative* use of both numbers and simple geometrical figures was widespread from time immemorial and constitutes a worldwide cultural phenomenon (see, for example, Andy Gregory's Chapter 2 on Pythagoreanism in this volume).

Second, to a non-existent historical route: within Western culture, which is our focus here, the usual chronological sequence that presents the early Church Fathers (Clement of Alexandria, Origen, Irenaeus) and Latin writers (Tertullian, Ambrose, Augustine, Cassiodorus) as direct heirs to the classical Greek tradition(s) of Pythagoreanism,[2] is historically unfounded. The few small tracts with proper number-symbolical contents passed on to us from Antiquity date in fact from no earlier than the late Hellenistic period (i.e. from the second to fourth century AD). These booklets are not just lacking in bulk but also in speculative density, and are more summaries of number symbolism than detailed expositions on the subject.

The Christian interpretation of arithmological data retains, up to a point, the three-fold pattern of ancient Greek culture regarding the interpretation of numbers in cosmological, ethical, and theological terms. In this manner, and according to context, any number may refer to a cosmological reality – there can be four elements, five senses, seven planets. On the other hand one or more numbers can represent various moral virtues: 4 steadfastness, 6 justice and balance, 8 harmony and perfection. A number can also point to a divine entity, whose proper nature and power they supposedly compound: 1 is equated with Zeus and Jupiter, 2 with Hera and Juno, 7 with Pallas and Athena, and so on. However, and perhaps as expected, the Christian interpreters were predominantly concerned with Scripture and its contents (as is the case with the Jewish author Philo of Alexandria in the 1st century AD) and, accordingly they took into account the manifold numerical instances present in the biblical texts. These far outreach the Decad (number 10) which was sacred to the Greeks, who, as such, declined to consider numbers above 10, with the eventual exception of 12 (Zodiac, Olympian gods) which frequently denotes completion.

The Bible was considered by Christian interpreters of number symbolism as a revealed or divinely inspired writing. This emphasized their core belief that any number mentioned in the Bible must necessarily conceal, beyond its literal aspect and its mere quantitative value, a hidden spiritual significance.[3]

Christian hermeneutics

Given these basic tenets, it becomes clear that the Renaissance or Early Modern period inherited the practice of number symbolism as a hermeneutical tool, designed mainly to interpret Scripture and shed light on the network of spiritual and messianic analogies and correspondences assumed to be linking both Testaments.

Combined with the three-fold exegetical pattern of cosmology, ethics, and theology mentioned above as regards pagan arithmological traditions, there came into existence another, again three-fold and this time doctrinal, Christian interpretation of number during the Antiquity and the Middle Ages.

The first of this new three-fold Christian interpretation is the professed divinely inspired nature of the Bible, which ensures the existence of a spiritual, non-quantitative meaning for each numerical instance occurring in the text. The fundamental idea being, as stated above, that nothing can be simply casual, much less devoid of a particular meaning, in the revealed Text. The second aspect is the sacred authority ascribed to the Church writers and the hermeneutical traditions manifested in their works. The third is the taking into account of the *ratio numerorum*, in other words the discursive analysis of the mathematical properties ascribed to a given number. In certain cases, the *calculus*, or basic arithmetical operations to which numbers may be subjected, also plays a significant part in determining their mystical value. For instance, the number 15 is sometimes considered as denoting the five senses, sanctified by the Trinity (5×3) and 12 can indicate the spreading of the Trinitarian Gospel to the four corners of the world (3×4).

As with symbols in general, numbers are generally regarded as admitting a plurality of meanings, according to their literary context and the theoretical perspectives they are embedded in. In other words, no numerical symbol may be considered as holding a given, definite, once-and-for-all mystical significance. On the contrary, any number may (and usually does) assume different meanings pertaining to different orders of reality and/or points of view. Augustine provides no less than 4 different interpretations of the number 153, the number of fish Simon Peter nets in John's gospel 21:11, and there exist at least some 14 others.[4] In this manner, whether quantitative value is in a privileged position over the significance supposedly imparted by its scriptural context, or the other way around, therein would lie the interpretation of a number. There is also a question of focus on Christian tenets or the secular knowledge in the crux of arithmological literature in the medieval period, the more so as common knowledge increasingly included, at the time, a degree of mathematical awareness.[5]

Formal number

The Platonic Renaissance brought about by the erudite enthusiasm of Italian Humanists took root in Florence, Venice, and Padua during the second half of the fifteenth century. Marsilio Ficino[6] (1433–99) was certainly the best-known exponent of this foundational current of Humanistic thought and resorted to number symbolism in several of his works and commentaries.[7] However, it is really his two younger contemporaries, the young Count Giovanni Pico della Mirandola (1463–94) and the German philologist Johann Reuchlin (1455–1522), who were in fact responsible for attracting a renewed attention to arithmology and turning it into a central feature of the Humanistic outlook.[8] Although neither of them wrote an actual treatise devoted to the topic in question, they indisputably succeeded in reinstating number symbolism as a major, almost autonomous, current within Renaissance Platonism, and within the domain of what is nowadays referred to as modern Western Esotericism. Pico is considered the founder of Christian kabbalah (an adaptation of Jewish-kabbalistic materials to Christian doctrines) and wrote extensively on the relations between magic, kabbalah, theology, and natural philosophy. Reuchlin – who met Pico in Italy – made a reputation for himself discussing the magical and spiritual powers of the divine Name, and became another major exponent of Christian kabbalah.

Pico claimed in his Roman writings of 1486–7[9] to be the first author to have restored its former intellectual status to the topic of symbolic numbers. He heralds the subject as a new *institutio philosophandi*, that is to say a way or a method of doing philosophy by way of numbers. In other words, this means that symbolism of numbers is no longer merely an allegorical treatment of mathematics, nor a hermeneutical tool designed for the exegesis of numerical passages in the Bible. It instead becomes a speculative trend of its own, an independent approach to philosophy, on a par with other currents such as Neoplatonism, magic, kabbalah, and so on.[10]

Such a method, presented by Pico as new, but 'which is in fact old', by his own admission in the *Oration*, was also, according to him, very much in favour with Plato himself and the ancient theologians. It is also, on the other hand, an autonomous 'art of number', supposedly capable of addressing problems of cosmology, natural philosophy, theology or metaphysics.[11]

In his *Conclusiones* (or 900 Theses) and *Apology*, Pico introduces the category of 'formal number' which entails, in a typical Neoplatonic fashion, that mathematical entities actually occupy an ontology of their own (and possess a specific mode of existence). Pertaining to an intermediary ('formal', i.e. intellectual) level of reality, and superior to the physical plane, numbers are, in this system of thought, endowed with a faculty of bestowing 'a power and an efficacy' on natural things, which belong in turn to the material level of existence.

In this manner, Pico relies much less on hackneyed numerical analogies than on the close interaction of the assumed cognitive and ontological properties of arithmetic. Number is thus a secondary cause exerting influence on the material world.

During his lifetime, handsome and rich Pico was sometimes called the 'Phoenix of his Age'. He was greatly admired but was also the object of much envy, which may account for his untimely death, probably by poisoning, at the age of 31.[12]

Even though he does not make use of the expression 'formal number', Reuchlin expresses quite similar views[13] in assimilating the essence of *intelligible number* with that of the Divine Intellect. From their conjunction, the primordial Unity, number flows – according to a process of emanation – into its own realm of being which, in turn, goes on to inform the material reality. Reuchlin, a vocal opponent of the destruction of Jewish books and a stout defender of the relevance of the study of Jewish religious culture to Christian scholars, explicitly called for a restoration of Pythagoreanism, which he considered to be an offshoot of Jewish kabbalah. In this respect, he derived this opinion from Philo and a number of late pagan thinkers and early Church Fathers, who asserted that the Greeks had borrowed their philosophy from the Jews and from Scripture. Pico stated, in his *Oration*, that as far as philosophy was concerned, he believed he found Plato and Pythagoras on every page of the kabbalistic works he read.[14] Such speculations obviously underlined and supported the theme of a close interaction between number symbolism, natural philosophy, theology, magic, and kabbalah.

The growth of an intellectual trend

In the wake of Ficino, Pico, and Reuchlin, and obviously facilitated by the development of the printing industry, the body of literature devoted to arithmology enjoyed a rapid growth from the sixteenth to the eighteenth centuries and beyond. In its spread, this literature gave birth to specific sub-genres and currents, and these are represented by books which have sometimes become reference works in this area. Let us look at some of these.

An essentially speculative current is that represented by some French disciples and friends of Jacques Lefèvre d'Etaples (?1460–1536) who, between 1510/11 and 1521, published in Paris several books on number symbolism. A celebrated French Humanist, Lefèvre d'Etaples first travelled to Italy and other European countries at the beginning of the 1490s, where he met or corresponded with M. Ficino, J. Reuchlin, J. Trithemius, and H.-C. Agrippa (1486–1535) (see Figure 6.1). His early interest in practical aspects of magic and number symbolism (which he combined in an unpublished treatise on *Natural Magic* (*c.* 1494)[15]) waned rapidly and he went back to the teaching of philosophy in Paris and to editing Aristotelian texts, scientific tracts (on mathematics, astronomy, and music, often in collaboration with members of his erudite circle, such as Bovelles or Clichtove) and mystical works by Christian authors (like those of Cardinal Nicholas of Cusa (1401–64)). This work came notwithstanding his biblical translations and personal involvement within the *Groupe(s) de Meaux*, an evangelical group, in which he played a role of an important religious reformer alongside Bishop Guillaume Briçonnet (1470–1534).

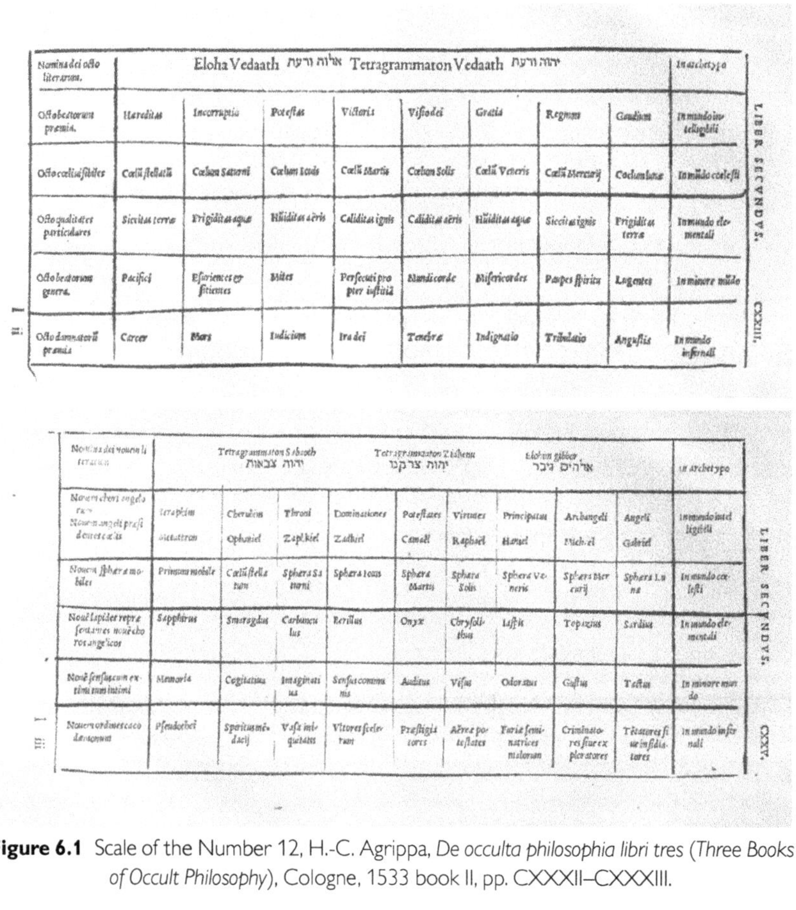

Nomina dei octo literarum.	Eloha Vedaath אלוה ודעת Tetragrammaton Vedaath יהוה ודעת							In archetypo
Octo beatorum premia.	Hæreditas	Incorruptio	Potestas	Victoria	Visio dei	Gratia	Regnum	Gaudium	In mundo intelligibili
Octo cœli visibiles	Cœlū stellatū	Cœlum Saturni	Cœlum Iouis	Cœlū Martis	Cœlum Solis	Cœlū Veneris	Cœlū Mercurij	Cœlum lunæ	In mūdo cœlesti
Octo qualitates particulares	Siccitas terræ	Frigiditas aquæ	Hūiditas aëris	Caliditas ignis	Caliditas aëris	Hūiditas aquæ	Siccitas ignis	Frigiditas terræ	In mundo elementali
Octo beatorum genera.	Pacifici	Esurientes & sitientes	Mites	Persecuti propter iustitiā	Mundicorde	Misericordes	Paupes spiritu	Lugentes	In minore mūdo
Octo damnatorū premia	Carcer	Mors	Iudicium	Ira dei	Tenebræ	Indignatio	Tribulatio	Angustia	In mundo infernali

Nomina dei nouem literarum	Tetragrammaton Sabaoth יהוה צבאות			Tetragrammaton Zidkenu יהוה צדקנו			Elohim gibbor אלהים גבור			In archetypo
Nouem chori angelorum / Nouem angeli præsidentes cœlis	Seraphim / Metattron	Cherubim / Ophaniel	Throni / Zaphkiel	Dominationes / Zadkiel	Potestates / Camaël	Virtutes / Raphaël	Principatus / Haniel	Archangeli / Michael	Angeli / Gabriel	In mundo intelligibili
Nouem sphæræ mobiles	Primum mobile	Cœlū stellatum	Sphæra Saturni	Sphæra Iouis	Sphæra Martis	Sphæra Solis	Sphæra Veneris	Sphæra Mercurij	Sphæra Lunæ	In mundo cœlesti
Nouē lapides representantes nouē choros angelicos	Sapphirus	Smaragdus	Carbunculus	Berillus	Onyx	Chrysolithus	Iaspis	Topazius	Sardius	In mundo elementali
Nouē sensus cum exterioribus interiores	Memoria	Cogitatiua	Imaginatiua	Sensus communis	Auditus	Visus	Odoratus	Gustus	Tactus	In minore mundo
Nouem ordines cacodæmonum	Pseudothei	Spiritus mēdacij	Vasa iniquitatis	Vltores scelerum	Præstigiatores	Aëreæ potestates	Furiæ seminatrices malorum	Criminatores siue exploratores	Tētatores siue insidiatores	In mundo infernali

Figure 6.1 Scale of the Number 12, H.-C. Agrippa, *De occulta philosophia libri tres* (*Three Books of Occult Philosophy*), Cologne, 1533 book II, pp. CXXXII–CXXXIII.

Rediscovered during the twentieth century, Charles de Bovelles (1479–1566/7) was an important philosopher and metaphysician, who enjoyed great fame in his lifetime and published several theological books, influenced by Pseudo-Dionysius the Areopagite (sixth century AD) and the work of Nicholas of Cusa.[16] As a Canon of Noyon Cathedral (in the North of France), he devoted his life, from 1515 onwards, to a peaceful religious life and to his studies. His interest in mathematics and geometry led to technical endeavours[17] as well as to a Pythagorean approach to number, relying heavily on analogy and correspondences. This is illustrated by his philosophical exposition of the ontology of number and the correlated ascension of the soul towards the Godhead in his *Liber de duodecim numeris* (Book of [the first] twelve numbers) and supported by geometrical considerations indebted to some extent to Nicolas of Cusa's famous *Learned Ignorance* (1440). Even though the overall perspective of the Book of twelve numbers is naturally Christian, critics have noted the puzzling scarcity of scriptural references in all but its last chapter (devoted to the number 12).

Our third example, Josse Clichtove (?1472–1543) (see Figure 6.2), was mainly a theologian but also taught philosophy in Paris and later became a well-known Humanist and Catholic reformer and was noted for his criticism of Lutheranism. Born in Nieuwport, he died in Chartres, leaving some important works of controversy and scriptural exegesis. His early booklet devoted to the theological hermeneutics of biblical numerals *Opusculum de mystica significatione numerorum* (Short treatise on the mystical meaning of numbers [contained in the Bible])[18] constitutes an erudite and mystically inclined summary of the patristic and medieval traditions regarding the interpretation of numbers within the Bible.

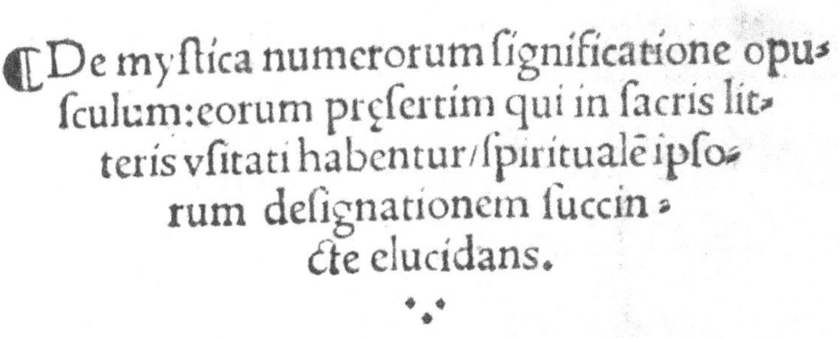
¶De myſtica numerorum ſignificatione opu-
ſculum: eorum prẹſertim qui in ſacris lit-
teris vſitati habentur/ſpiritualē ipſo-
rum deſignationem ſuccin-
cte elucidans.

Figure 6.2 Title page of J. Clichtove, *De mystica numerorum significatione opusculum*, Paris, H. Estienne, 1513.

A member of the same circle, Gérard Roussel (?–?) was, like several other French Humanists, a protégé of the sister of the King Francis I, Marguerite of Angoulême (1492–1549). He became Abbot of Clairac, then Bishop of Oloron (*c.* 1538–40), before probably converting to Protestantism. His erudite arithmological commentary on Boethius' *De Arithmetica* (Manual of Arithmetic)[19] is a rather uncanny attempt to correlate the quantitative and qualitative understandings of number.

In all three texts, in typically Neoplatonic fashion, and contrary to Cusa's perspective, the status of arithmetic is considered paramount and superior to that of geometry. Even for Bovelles, 'arithmetic is devoted to numbers, which lay and are posited in the soul', whereas geometry is more intrinsically concerned with material and corporeal measures, quantities and dimensions.

These volumes demonstrate the continuity, into the era of the printing press, of many different trends in the practice of arithmology, namely

1. The serial examination of the primary integers, of their mathematical structure, individual properties, and mutual relations, and of their varied analogies and correspondences in the realm of cosmology and metaphysics (with, surprisingly in Bovelles' case, almost no reference to scriptural contents).
2. The strictly orthodox and theological interpretation by Clichtove of numerical statements in the Bible, relying exclusively on a library of Church Fathers and Catholic authors.
3. A learned, mystically inclined commentary (Roussel) on a famous scientific and philosophical manual, the contents of which are altogether devoid in themselves of any number-symbolical leanings, yet allow for a theological transposition of number.

Seen from this perspective, arithmetic constitutes an approach towards divine realities which, in turn, provide the foundation for the existence of the mathematical realm.

Universal knowledge

Another distinct category of arithmological texts is that of the encyclopaedias or lexicons that attempt to present a detailed account of number symbolism within a framework of a spiritual, erudite, and didactic endeavour, inspired by the cultural atmosphere of the Counter Reformation. Such treatises mobilize vast resources of learning in order to establish the desired *unison* of secular knowledge, Scripture, and Roman Catholic doctrines, through the mediation of the analogical transpositions of number.

Pietro Bongo's (?–1601) *Numerorum Mysteria* (Mysteries of Numbers)[20] and Athanasius Kircher's (1601–80) *Arithmologia*[21] both represent the most well-known examples of such efforts, which have largely contributed to the establishment and legitimization of the cultural relevance of this topic within Western culture. As such these works give number symbolism a wide intellectual appeal.

A member of an ancient and noble family from Bergamo, as well as a canon of the local Sant'Alessandro Cathedral, the immensely erudite P. Bongo wrote a 700-page quarto tome on Pythagoreanism, drawing from every conceivable source available to him at the time, including magical, hermetic, alchemical, and esoteric writings (Pico, Ficino, Dee, etc.). Although he was sometimes criticized for the heterogeneous character of his book, which takes numbers into the hundreds of thousands and sometimes degenerates into a mere dictionary of the allegorical meanings of numbers (a lexicon which members of almost all artistic or intellectual trades could indifferently tap into), Bongo remained aware of certain theoretical issues of his time. The growing opposition between traditional thinking, which was considered as authoritative and mainly based on analogy, correspondences, and the powers hidden in the essence of things, and the modern thought, which attempted to read the world through the lenses of experience and of a quantitative, discursive mathematical language, was not lost on him. Bongo contrasted the Book of Nature with that of Scripture, and exhibited an understanding of Pythagoreanism as an intermediate hermeneutical device useful in the spiritual interpretation of the Bible, and of culture, with pastoral aims. Following this perspective, he naturally retained a mainly theological outlook, which included the allegorical use of numbers as a general key of the interpretation of reality.

Another interesting figure in this respect is the Jesuit Father Athanasius Kircher (see also Chapter 5 by Wilson and Fauvel in this volume). He was a famous polymath, with an almost universal interest in ancient languages and philosophical, scientific, occult, and historical issues. Within his enormous literary output, the *Arithmologia* is the only volume exclusively devoted (in six parts) to different aspects of number symbolism, including divination, magic squares, astrology, the magical and talismanic uses of letter symbolism and its numerical applications. Only the last chapter is actually devoted to arithmology proper. Although cautious to remain within the frame of a strict religious orthodoxy, Kircher's well-known fascination with the topics of magic, occult sciences, and the mysteries of number remain nevertheless palpable throughout his work. His aim, in this final part of his treatise, is essentially to expose a Christian philosophy of number, in which its intelligible nature is shown to pervade the entire creation, so that the 'mystical reasons' of number reveal the hidden harmonies of nature as well as their source and analogies within the divine realm.

Although much less encyclopaedic in scope, one may also take the opportunity of mentioning the first printed English tract on arithmology, William Ingpen's *The Secrets of Numbers*.[22] The book contains some 20 short chapters evoking the Decad and a few greater digits. It also describes instances of the cosmological, scriptural, and theological symbolism of number, without any claim to originality but with a good knowledge, nevertheless, of earlier arithmological material.

Similar endeavours, but trimmed down to the study of a single, particular number, also exist and, like the preceding titles, were mainly published at the end of the sixteenth century or early seventeenth century. In these works the main focus is usually on numbers 3 and 7, considered as being of outstanding importance within the first 10 or 12 natural integers, and as a synthesis not only of the Pythagorean Decad, but of the whole contents of the universe. Most significant in this respect is Alessandro Farra's (?–after 1577) *Settenario dell'humana riduttione* (Septenary of human reduction[23]) which deals with the seven degrees or steps of the mystical ascension, leading to the acquisition of divine wisdom along a spiritual path marked by seven phases related to seven mythological figures, from Mercury to Orpheus. In passing, Farra also makes use of the category of *formal* number, inherited from Pico. The last section of the book, dealing with the definitive possession of wisdom and the contemplation of divine archetypes, is devoted to what the author calls 'symbolic philosophy, or the emblems'. These 'emblems' represent the uppermost mode of communication between man and the Godhead (the *Settenario* is also very much about rhetoric), akin to mystic silence. It is this part of the treatise that considers Pythagorean numbers, alongside classical erudition and Christian-kabbalistic materials. The number 7 features here by description of its synthesis from its main elements, 3 and 4, as well as by explanation of the corresponding geometrical figures.

Centring on classical erudition and the magical effects of poetry, such as those traditionally attributed to Orpheus by Greek mythology, Fabio Paolini's (1535–1605) *Hebdomades, sive septem de septenario libri* (Hebdomads, or seven books on the septenary[24]) devoted 49 (!) chapters to the manifold interpretations of one single verse by Virgil,[25] as well as to the symbolism of the musical scale. The book's purpose was to study the seven-fold way of correlating music and rhetoric to natural and celestial magic. This the author did with an aim of recreating the achievements of Orpheus himself, who supposedly built cities and charmed living beings by means of his poetry and music alone. According to this perspective, the septenary structure presents a typology of celestial influences and of the categories of poetry and musical practice, thus providing a living and supposedly efficient link between the two. Strange as they may seem to us, such purposes and learned discussions were in fact relatively common in Italian academic and artistic circles of the late sixteenth and early seventeenth centuries, where the idea of retrieving the wonder-working powers ascribed to the art made by the ancients was very much favoured and was considered a real possibility. While certainly not the only treatises devoted to the numbers 3 and 7 in print during the seventeenth century[26] and later periods, Farra's and Paolini's volumes provide representative examples of their form and content.

Science and theology

Some well-known textbooks about mathematics, albeit written essentially to serve as scientific statements, also mention the qualitative uses and analogies of number.

In the first, more theoretical, part of his *De divina proportione* (On the divine proportion[27]) Luca Pacioli (*c.* 1445–1517) studied at length the mean and extreme ratio (also commonly known as golden section) and its supposed role in the creation of the material world.

Basing himself on Plato, Euclid, and Vitruvius, Pacioli links certain characteristics of the five Platonic solids (tetrahedron, cube, octahedron, icosahedron, and dodecahedron) from the *Timaeus*, and of the golden mean, to five divine attributes (unity, trinity, transcendence, immutability, and creative wisdom); see Figure 6.3. He thus extolls the importance and perfection of this particular proportion, of the number 5 and of the five regular polyhedra within the fabric of the universe. Written in Italian, the book aims to communicate, after the preliminary theological discourse on divine and cosmic harmony, some elements of scientifically grounded geometrical knowledge to the Renaissance architects, stone-cutters, and sculptors whom Pacioli frequently visited and taught.

Pacioli was a renowned professor of mathematics and author of several classic manuals of algebra, chess, and accountancy. Having studied and become a Franciscan friar in Venice (around 1476), he afterwards travelled throughout Italy in order to teach mathematics and the *Elements* of Euclid (in Perugia, Florence, Naples). He met the famous architect and painter L.-B. Alberti (1404–72) in Rome and, later, in 1496, met Leonardo da Vinci (1452–1519) in Milan. Da

Figure 6.3 Five platonic solids: tetrahedron, octahedron, cube (hexahedron), dodecahedron, and icosahedron.

Vinci became his friend and contributed a famous series of engravings representing some 60 complex geometrical bodies to the above-mentioned *De divina proportione*. Pacioli translated Euclid into Latin in 1509 (and possibly into Italian as well – however this version is lost), and inherited some works on perspective and on the Platonic bodies from his friend the celebrated painter Piero della Francesca (*c.* 1420–92). It seems that some of della Francesca's work was later included in the *Divine Proportion* without giving the credit to the original author. Pacioli left Milan with Leonardo in 1499 and went with him to Mantua, then later to Bologna and Rome, where he died.

Of a different tone, but exhibiting a similar conceptual continuity between the theoretical and the applied aspects of mathematics, John Dee's (1527–1608/9) famous *Mathematicall praeface* to the first English translation of Euclid's *Elements*[28] praises the 'noble science mechanicall, and practitioners thereof', while describing no less than 19 different branches (methodicall arts) of mathematical science. For him as well, mathematics constitutes the essence of reality and to understand mathematics is to understand the law of creation and the structure of nature. The magical power of numerical entities reside in the fact that they derive from the divine Monad (unit) and link up the material, celestial, and divine worlds, while having the power to influence nature.[29] Yet, similar to man himself, number is for Dee a central feature within the universe, encompassing all levels of creation and mediating between the divine and material spheres. Mathematics, therefore, since it operates on all these planes, may serve to implement an ambitious synthesis of science, magic, kabbalah, and natural philosophy. It sits within the theoretical foundations of different arts and crafts, deciphering the secrets of nature and, in so doing, revealing the divine mysteries as well. These were revelations which Dr Dee desperately sought in his lifetime through angelic agency and scrying. He practised both from 1581 to the end of his life, with the help of various seers, in England as well as on the Continent, where he resided between 1583 and 1589.

A leading scientist of the Elizabethan Renaissance, having travelled in Europe and taught mathematics in Paris in his early 20s, Dee was in relative favour with Queen Elizabeth I (1533–1603) to whom he sometimes acted as counsellor. Dee had a famously large private library and collection of scientific instruments (later dispersed), and was known as an alchemist, mathematician, geographer, astronomer, and magus. His versatility was relatively common for learned men of the time who worked in an age before the disciplinary boundaries of modern scientific disciplines were firmly established. Dee, however, was also part of a long line of humanist thinkers who were not only preoccupied with scholarly sources and an exclusively theoretical approach to knowledge but who also, like Pacioli, Bovelles, and the Frenchman Girard Desargues (1591–1661), maintained an interest in underlining or even developing the technical applications of algebra and geometry. In so doing, they intended to put contemporary mathematical discoveries at the immediate disposal of craftsmen and artisans alike. Their aim was to provide them with the scientific background necessary to shed light on empirical procedures, mainly hitherto handed down by secretive professional traditions. Such a tendency, which was subsequently greatly expanded in the seventeenth and eighteenth centuries, and meant a democratization of knowledge and of its accessibility, also aimed at ending the clerics' monopoly on learning in general.

As we have briefly shown, Renaissance Humanism has, among many other achievements, enabled the emergence of a specific body of literature dedicated to number symbolism, de facto enabling Pythagoreanism as a way of numerical interpretations to secure a niche of its own within both European culture and Western esoteric currents.

Such an intellectual trend about number and geometry is almost entirely supported by a Neoplatonic worldview, which tends to ground the study of nature and cosmology in ontological and theological perspectives. Like most other esoteric currents of the Early Modern period, Pythagorean arithmology implies the conception of number as a formal cause. This is even more so in the Christian Neoplatonism which, ever since Augustine, tends to identify the archetypal essences of things as numerical entities in the Divine Intellect, which served as models for the creation of the universe.[30] Such a concept is connected to a doctrine of the unicity of creation, that is a network of cosmic analogies and correspondences pervading and linking the hierarchy of different planes of existence, composing the universe, and construed as an organic, living being. The progressive decline of such a worldview within European culture over the course of the seventeenth and eighteenth centuries seriously damaged the perceived relevance of arithmology as an intellectual tenet.

Increasingly, as a consequence of this decline, the concept of a meaning attached to number in the way discussed in this chapter was reduced to the dimension of a mere logical tool, useful for measuring and effectively describing physical reality but no longer plausibly referring to its supposed inner life or essence. This therefore rendered meaningless number's former mediation between the spiritual and material dimensions of creation, as well as its role in transmitting supernal influences. The gradual relinquishment of the quest for patterns of structural elements, perceived to be emanating from the mind of God and as supposedly expressed by the mystic harmonies of numbers and geometrical proportions, were lost in mathematics that subsequently developed.

Notes and references

1. The term 'number symbolism' did not exist under that name during the Renaissance or the earlier periods of Western cultural history. 'Number symbolism' or arithmology are relative neologisms. The Latin tradition preferred to term the relevant discipline 'mystical arithmetic', *mysterium* (or: *sacramentum*) *numerorum*, sometimes *mysticus numeri sensus* or *mystica numerorum applicatio* – *symbolum* being somewhat rare (mostly applied to number) during the Middle Ages (the more common would have been *figura*, *exemplum*, *typus*, etc.).
2. Christoph Riedweg, *Pythagoras. His Life, Teaching, and Influence*, Cornell University Press, 2005.
3. For a contemporary assessment of these views, see John J. Davis, *Biblical Numerology. A Basic Study of the Use of Numbers in the Bible*, Baker Book House, 1968 (reprint 1985).

4. Davis, *Biblical Numerology*, p. 147.

5. As with the so-called 'Latin Platonists': Macrobius (fifth c.), Martianus Capella (fifth c.), Chalcidius (fourth c.), Boethius (fifth–sixth c.), whose works constitute the backbone of Western erudition until well into the Middle Ages. The two last named authors were Christian. J.-P. Brach, 'Arithmology', in K. Pollman and W. Otten (eds), *The Oxford Guide to the Historical Reception of Augustine*, Oxford University Press, 2013, III, pp. 73–6.

6. A Roman Catholic priest and a Canon of San Marco in Florence, Ficino published (1484) – thanks to the sponsorship of the then ruling Medici family – the first integral translation of Plato's works in Latin, as well as that of many other important ancient Greek philosophical and magical sources. The impact of his work on Renaissance and later European culture cannot be overstated.

7. Michael J.B. Allen, *Nuptial Arithmetic: Marsilio Ficino's Commentary on the Fatal Number in Book VIII of Plato's Republic*, University of California Press, 1994.

8. Along with the Benedictine Abbot of Sponheim, and later Würzburg, Johann Trithemius (1462–1516); Noel L. Brann, *Trithemius and Magical Theology. A Chapter in the Controversy over Occult Studies in Early Modern Europe*, SUNY Press, 1999, 114 ss. Some of Trithemius' letters on the virtues of numbers were destined to Germain de Ganay (?–1520), who also corresponded with Bovelles (see below) on the subject of arithmology.

9. *Oration on the Dignity of Man*, ed. and trans. F. Borghesi, M. Papio, and M. Riva, Cambridge University Press, 2012; *Conclusiones* (or *900 Theses*) ed. and trans. S. Farmer, *Syncretism in the West: Pico's 900 Theses (1486): The Evolution of Traditional Religious and Philosophical Systems*, MRTS, 1998; *Apologia*, ed. and trans. P.E. Fornaciari, *Apologia. L'autodifesa di Pico di fronte al Tribunale dell'Inquisizione*, Sismel, 2010.

10. *Oration*, <http://vserver1.cscs.lsa.umich.edu/~crshalizi/Mirandola/>, section 32 (retrieved 09/01/2014).

11. A section of the *Conclusiones* lists no less than 74 different questions which the author boasts to be able to answer through the use of number.

12. M. Moore, 'Medici philosopher's mystery death is solved', *The Daily Telegraph* (London; 07/02 2008); <http://www.telegraph.co.uk/news/worldnews/1577958/Medici-philosophers-mystery-death-is-solved.html> (retrieved 15/01/2014).

13. *De arte cabalistica* (1517); *On the Art of the Kabbalah*, trans. by M. Goodman and S. Goodman, Nebraska University Press, 1983.

14. *Oration*, <http://vserver1.cscs.lsa.umich.edu/~crshalizi/Mirandola/>, section 35.

15. J.-M. Mandosio, 'Le *De Magia naturali* de Jacques Lefèvre d'Etaples. Magie, alchimie et cabale', R. Gorris Camos (éd.), *Les Muses secrètes. Kabbale, alchimie et littérature à la Renaissance. Hommage à François Secret*, Droz, 2014, pp. 37–79.

16. Joseph M. Victor, *Charles de Bovelles, 1479–1553: An Intellectual Biography*, Droz, 1978.

17. Among which the first textbook of geometry actually printed in French (1511). According to some of his contemporaries, he was often seen meditating in front of geometrical figures pasted on an opposite tree or wall.

18. Published in Paris in 1513 *Opusculum de mystica significatione numerorum* is the first printed title exclusively devoted to the subject of arithmology proper.

19. G. Roussel's commentary on Boethius' famous textbook (where he follows the critical edition of the text procured earlier by Lefèvre d'Etaples himself) was published in 1521.

20. Bergamo, 1583–85 and – augmented – 1599 (there are several other printings; reprint Olms Verlag, 1983 – with an important introduction by U. Ernst). G. Piccinini, L'opera di Pietro Bongo sulla simbologia dei numeri, *Archivio Storico Bergamasco*, **6**, 1984, 105–11.

21. Published Rome, 1665. See P.K. Findlen (ed.), *Athanasius Kircher: The Last Man Who Knew Everything*, Routledge, 2004.

22. Published London, 1624; reprint BiblioBazaar, 2010.

23. Published 1571 (2nd edn 1594).

24. Published 1589.

25. *Aen.* VI, 646: *Obloquitur numeris septem discrimina uocum* («[here Orpheus too, the long-robed priest of Thrace] accompanies their voices with the seven-note scale »).

26. They are referenced in our study: Mathematical esotericism: some perspectives on Renaissance arithmology in W.J. Hanegraaff and J. Pijnenburg (eds), *Hermes and the Academy. Ten Years' Study of Western Esotericism at the University of Amsterdam*, University of Amsterdam Press, 2009, pp. 75–89, on which the present article draws in part.

27. Venice, 1509 (but written in Milano between 1496/8); E. Giusti and C. Maccagni (eds), *Luca Pacioli e la matematica del Rinascimento*, Giunti, 1994.

28. London, 1570 (facsimile reprint by A.G. Debus, New York, Science History Publications, 1975). Stephen Clucas (ed.), *John Dee: Interdisciplinary Studies in Renaissance Thought*, Springer, 2006.

29. This is of course quite reminiscent of Pico's 'formal number', an expression which Dee does acknowledge; cf. J.-M. Mandosio, Beyond Pico della Mirandola: John Dee's "formal numbers" and "real cabala", in J. Rampling (ed.), John Dee and the sciences: early modern networks of knowledge, *Studies in History and Philosophy of Science*, **43**(3), 2012, 489–97.

30. Aug., *De Trinitate* ('On the Trinity'), VI, X, 11.

CHAPTER 7

Newton, God, and the mathematics of the two books

ROB ILIFFE

Isaac Newton's wide range of intellectual interests has received extensive attention over the last few decades. The recent publication of his non-scientific writings means that historians can now address his work in theology and alchemy, in addition to his more famous achievements in the exact sciences. As a result, it is possible to assess the ways in which his religious beliefs informed his views in natural philosophy (or vice versa), and to compare his research on prophecy and church history with his work in natural philosophy. The subject of mathematics presents an interesting test case for exploring Newton's work in these areas since, as he saw it, both the natural world and Scripture were written in a mathematical language. In this chapter, I address two key questions. To what degree did his adherence to the tradition of natural theology rest on a conception of God's mathematical skill, inherent in the creation of a world governed by mathematical laws? To what extent, if at all, did Newton use his special mathematical prowess in understanding the numbers that appear in both Old and New Testament prophecy?

Although Newton's contemporaries were aware that he was a devout man who knew the Bible as well as anyone, when he died in 1727 there was little published evidence of his religious views. Soon after his death, two works appeared indicating that he had undertaken serious research into prophecy and into the ancient history of the world. However, he left almost no traces of his deeper religious interests in the first editions of his great works *Principia Mathematica* (1687, 1713, and 1726) and *Opticks* (1704, 1706, and 1717/18). Under some pressure to reveal these views, in the second and subsequent editions of these works he added some pertinent remarks on how the harmonious order of the cosmos indicated that it had been designed and brought into being by an intelligent deity. He also included some important metaphysical statements on the nature of God, and on the relationship between God and man.[1]

This was the public face of Newton's religiosity. However, a handful of people were privy to the fact that there was another facet to his religious beliefs, and rumours about these views were in circulation during his lifetime. The manuscript record shows that early in his career, he developed an intense hatred of the orthodox belief that Jesus Christ was 'co-equal' to and 'consubstantial' with the Father. He had to mask this heretical view from his colleagues, since the doctrine of a holy and undivided Trinity was a central tenet of the Church of England to which he publicly professed allegiance (and the phrase was enshrined in the name of his own college). Nevertheless, from some point in the 1670s, he came to understand that the doctrine was a key component in what he considered to be the diabolically corrupt religion introduced by the Western church in the fourth century after the birth of Christ. Numerous notes, orphaned chapters and full blown treatises attest to Newton's great efforts to discover the history of this great corruption of Christianity. As he saw it, the proper understanding of this process required the mastery both of church history and of the mysterious images that appeared in the prophetic books of the Old and New Testaments. 'Mystical' numbers, some repeated in both parts of the Bible, peppered these prophetic utterances and, like others, Newton believed that Christians had an obligation to understand them.[2]

The human intellect and the mathematical world

In the first two books of the *Principia Mathematica*, Newton derived mathematically from the three laws of motion a series of propositions about motion in resisting and non-resisting media. In Book Three, he showed that the motions and forces that made up 'the system of the world' were mathematically precise, and were governed by the law of universal gravitation. His contemporaries revered Newton and his achievement in part because they believed that he had revealed the mathematical foundations of a divinely created world. In his peculiarly Lucretian 'Ode' to the *Principia*, which appeared at the start of the work itself, Edmond Halley waxed lyrical on how the text laid out those laws 'which the all-producing Creator, when he was fashioning the first beginnings of things, did not wish to violate, and established as the foundations of his eternal work'. The clouds of ignorance over the causes of terrestrial and celestial phenomena had been dispelled by mathematics, and Halley exulted that 'the acuity of a sublime Intellect has allowed us to penetrate the dwellings of the gods and to scale the heights of heaven'. He famously concluded the poem by saying that no mortal could come closer to the gods. However, when he claimed that Newton's mind was 'possessed with the fullness of divinity', he was not engaging in mere hyperbole. Like many others, he was genuinely stunned that Newton had shown that human beings possessed an intellect that was adequate to grasping the fundamental mathematical structure of nature – if you like, the mind of God.[3]

In his published writings, Newton said very little about what this mathematically exact cosmos implied about the nature of its creator. In private, however, he said rather more. In a remarkable document probably composed soon after he had become Lucasian Professor in 1669, he launched a withering attack on the concepts of force, motion, and space that Descartes had articulated in his *Principia Philosophiae* of 1644. At one point, he argued that the Cartesian notion of 'place' was so incoherent that not even God could retrospectively calculate the original position of an object from its current location. Later on in the text, Newton took particular exception to the manner in which Descartes equated body with extension. According to this view, having removed all subjective qualities, the essence of a body was constituted by the volume bounded by its length, width, and height. Newton countered that extension could not by itself perform the tasks that a substance could – that is, it could not resist or in any other way interact with other bodies, it could not be perceived, and it would not disappear if God suddenly caused a body occupying the same space to be annihilated. Extension was not, however, *nothing*, and Newton proceeded to argue that empty space was divided up into an innumerable amount of really existing but invisible extended parts, all of which had the characteristics of mathematical figures such as spheres, cubes, and triangles. For example, space was spherical, not merely at those moments and in those places when and where it 'contained' a sphere that occupied various parts of space as it moved through it, but *at all times and everywhere*.[4]

Space extended infinitely far in all directions, Newton continued, because there was no limit in space beyond which we could not imagine a further space. Some mathematical figures (lines, paraboloids etc.) extended to infinity, and Newton emphasized that these objects were real and not merely imaginary. Indeed, only the understanding could positively grasp extended objects that were so distant as to go beyond the power of the imagination to conceive. Towards the end of his essay he made a novel attempt to explain how God might have created body or substance in the world. Perhaps God could make certain defined spaces impenetrable to other bodies, and these could be moved from one part of the universe to another according to certain laws. God could also constitute such spaces so that they could be experienced by sentient beings, and such an object would have all the required characteristics of a body – 'the product of the divine mind realised in a definite quantity of space'. Newton added that one of his goals was to show that God's faculties were somehow adumbrated in mortals. God's creative powers were mimicked in the capacity of humans to move their own bodies by the actions of their wills, though humans were subject to divine laws and could not create something from nothing. Newton also alluded to the fact that human beings possessed an intellect that resembled – to a finite degree – the infinite capacity of God's understanding. His own work was, in a sense, a test case of how far the divine mind could be accessed by the human brain.[5]

Newton contrasted the finite intellectual abilities of humans with the unlimited capacity of God's understanding in other unpublished writings. In one essay from early 1685, which contained many of the elements that would soon appear in the *Principia*, he included a brilliant paragraph in a scholium at the end of Theorem 4. He argued that whether the solar system was considered to be moving or at rest, the fact that its centre of gravity coincided exactly or, more usually, very nearly with the sun, was an a priori proof of the Copernican system. Because of mutual attraction between the heavenly bodies, the distance between the sun and the centre of gravity of the solar system was constantly changing. Hence, planets did not move in exact ellipses, and never traversed the same orbit twice. As a consequence, Newton went on, the paths carved out by planets as they moved through space were literally incalculable by mortals: 'unless I am mistaken, to consider all these causes of motion at the same time, and to define these motions by exact laws that permit ready calculation, is greater than the power of the entire human intellect'. He left it up to the reader to draw the obvious conclusion that the only being capable of making so many calculations was God himself. In determining the orbit of a particular planet, mortals did not have the calculating power of the deity. If they made only a few observations of the object at various points in its orbit, no matter how accurate and precise the measurements were, they would be unable to deduce an 'ideal' path for the body on account of the contingent gravitational forces that were exerted on the planet at those moments. Instead, Newton concluded, one could calculate the planet's approximately elliptical orbit by interpolating many 'mutually moderating' measurements of the orbiting object.[6]

Newton's reference to a perfect geometrical world created by God reappeared in a different form in the *Principia* itself. In the 'Preface to the Reader' that was found at the start of the first edition of the work, he made a series of foundational comments on the disciplinary nature of the book. At the start of the text he praised the moderns for attempting to reduce natural phenomena to mathematical laws. Nevertheless, he also offered an encomium to the approach taken by the Ancients, who had divided mechanics into the 'rational' and the 'practical', the former proceeding rigorously through demonstrations. Newton effectively equated geometry with rational mechanics, contrasting the precision inherent in geometry with the inexactitude of practical mechanics. In the following sentences he recast these definitions in operational terms, and noted that the imprecision inevitable in practical mechanics was due to human error, rather than because of problems in mechanics itself. If someone worked with perfect exactness, he noted, 'he would be the most perfect mechanic of all'. Although this phrase was expressed conditionally, Newton noted that rational mechanics was the discipline most proper for understanding *in exact terms* the phenomena and causes of a natural world created by a perfect mechanic. Geometry was part of mechanics precisely because it was the science of those real, mathematically exact motions and forces that had been fashioned by God.[7]

Intelligent design

The clearest articulation of Newton's belief that God was the author of a mathematically precise and well-ordered system came in a correspondence he conducted with Richard Bentley in 1692–3. Bentley, an ambitious classicist who would be master of Newton's own college within a decade, had given the first series of lectures that aimed to rebut atheism under the terms of a bequest left by Robert Boyle. Now preparing them for publication, he must have thought it strange that there was only one reference to God in the first edition of the *Principia*. He appealed to Newton for advice on how to use the doctrines of the

Principia to good effect against unbelievers, and Newton reassured him that while he was writing the work, he had borne in mind arguments that might reinforce belief in God among thoughtful individuals. Bentley asked Newton a series of probing questions, and Newton's replies included genuinely novel and powerful additions to the battery of arguments used by scholars who, working in the tradition of natural theology, inferred the existence and attributes of the deity from the present order of the world.[8]

It is important to realize that many of Newton's arguments were hypothetical, being based on certain assumptions that Bentley had raised in his initial (now lost) letter. Nevertheless, in his preliminary remarks, he gave reasons in favour of positions to which he was clearly committed, such as the existence of an infinite universe. He suggested that if all the matter in a *finite* universe had at first been evenly distributed across that universe, and each particle possessed 'innate' gravitational attraction to all other bodies, then over time, universal gravitation would have forced all these bodies to congeal into one giant body at the centre of the universe. However, if at the beginning of time all material particles had originally been located at equal distances from each other throughout an *infinite* cosmos, then over time, mutual gravitation would have given rise to 'an infinite number of great masses' placed across space at great distances from each other. Newton did not make it clear whether this argument was supposed to show that God would not have created matter with innate gravity in a finite universe, or whether we were to infer that the universe had always been infinite from the fact that we now saw large bodies widely distributed across the universe. He added that natural causes could not by themselves have produced the different species of celestial body, some very hot and others apparently lifeless, which now constituted the solar system. A complex planetary system designed so that one (and only one) sun was at the centre of six revolving planets, three of which had their own satellites, must have been designed by a voluntary agent who acted either by natural or supernatural means.[9]

Newton added that the motions of the planets also had to be 'imprest' in them at some stage by an intelligent, calculating being who was adept at mathematics and engineering. Although comets travelled in all sorts of directions, planets travelled in the same directions and in approximately the same plane – a feature that must have been the effect of design. No natural cause by itself could have produced the harmonious arrangement by which each planet and its satellites was endowed with the precise locations, masses, and velocities that it now had. Nor could it have given rise to the mathematically precise laws that governed their interactions. Indeed, if any of these values had been awry by even a small amount, then without some external intervention, the system would have become chaotic. Recalling the claim he had made in the draft of the *Principia*, Newton argued that making such a system work required a 'Cause' that understood and compared the masses, gravitational forces, interstitial distances, and velocities of all the bodies in the solar system. Only a being highly skilled in mechanics and geometry could 'compare & adjust' all these elements – and perhaps only a super-intelligent mortal could understand them. Divine foresight was nicely shown in Newton's favourite case of the gravitational interaction between Saturn and Jupiter. Having presumably been created before the solar system took its present form, they had been placed by the creator at the positions from the sun that they now occupied to prevent their mutual attractions from destabilizing the system.[10]

In reply, Bentley pressed Newton about exactly how his comments regarding the state of the solar system concerned the design argument. Underlying the question was the principle of minimizing the amount of work God could be said to have performed in the creative act. Bentley also pressed Newton on whether the current orbits of the planets could have arisen by the force of gravity alone, and in response Newton considered a situation where the earth was placed with neither gravitational force nor motion in the centre of the *orbis magnus*, that is at the centre of the terrestrial orbit around the sun. If it were then endowed with both gravitational attraction towards the sun and the correct inertial tangential motion, this would be sufficient for it to take up its recurrent orbit around the sun. However, Newton added that it would be impossible for the earth to acquire these motions by natural causes alone. He referred to an argument by Plato (allegedly in the *Timaeus*, and rehearsed by way of Galileo in François Blondel's *L'Art de jetter les Bombes* of 1683), in which the latter had supposedly suggested that the deity might have set up the current motion of the planets by creating them very far from their current positions and then letting them fall into the sun – at which point the deity would need to turn the falling motion into a transverse one. Newton noted that this would give rise to the current planetary orbits only if the gravitational power of the sun doubled at exactly the moment when each planet arrived at its proper distance from it. As a consequence, God would be required to perform two separate acts, simultaneously turning the motion of the falling planets sideways and doubling the power of the sun. It was much more in accordance with the notion of an intelligent deity to suppose that he had framed the current system of the world in one act.[11]

Holy root extractions

Based on the authority of the general claims made in the *Principia*, Newton's letters to Bentley greatly strengthened standard 'design' arguments. These were not, however, the only strings to his bow. At the end of his letter to Bentley of December 1692, he let slip that he had another argument in favour of the existence of God that was 'very strong', but added that 'till ye principles on wch it is grounded be better received I think it more advisable to let it sleep'. If this demonstration had pertained to natural theology, then Newton would not have hesitated to mention it. On the other hand, we know that the bulk of his life was spent working on religious matters, and that he believed that a proper understanding of Scripture and a prophetically informed sacred history provided proof of the existence of God and of his working throughout history. Newton may have let his other argument rest because he did not want anyone else to divine the true nature of his own anti-Trinitarian religious views. On the other hand, if he harboured suspicions that Bentley would not be favourable to arguments drawn from prophecy, they were well founded.[12]

The numbers embedded within prophetic texts had always attracted the attention of exegetes. For Christians, the most important numbers were found within Revelation. The text was initially shunned by leading Protestants, who feared that its contents encouraged social unrest and promoted chiliasm, a view that (according to its detractors) held that the future millennial reign of Christ would offer endless opportunities for hedonistic pleasures. By the middle of the century most Protestants understood that it offered a coded account of the battle between the true church and its false images. For those who were skilled interpreters, the numerical mysteries contained within its verses – which often drew on prophetic statements in the Old Testament – were keys to unlocking the nature of Antichrist. In some cases the faithful were explicitly enjoined to perform arithmetical operations. In his late second century *Adversus haereses*, for example, the church father Irenaeus had spent some time analysing the number 666 and had concluded that it might depict the name LATEINOS. This was because in the common Greek alphabet letters stood for specific numbers, and if one added up the numerical values for the letters in LATEINOS (according to the practice known as gematria) they totalled 666.[13]

Because it seemed to point to Rome as the chief abode of Antichrist, Irenaeus's argument, and many like it, became entrenched within the Protestant tradition as it developed in the sixteenth century. Certain numbers stood out as having special significance, and at times readers of the Bible were told to count, or measure with a reed, various features of mystical buildings. Some numbers were crucial to the Protestant tradition, and to the Christian tradition more generally. The number 7 occurred regularly in the first chapters of Revelation and indeed throughout the New Testament. It had special status as a mark of the godly, and successive seals, trumpets, and vials were taken to refer to a succession of the most significant historical events in Christian and Protestant history. The number 12 also had connotations of righteousness and godliness. It was the number of the tribes of Israel and of Jesus' disciples, and was repeated throughout Rev. 21 in the description of various dimensions of the city of the New Jerusalem. In Rev. 7 John was told that 12,000 people from each of the 12 tribes of Israel had been given the seal of God on their foreheads. The figure was apparently repeated in an obscure (squared) form in Rev. 14:1–5, where John saw the Lamb on Mount Zion with 144,000 redeemed virgins.[14]

Interpreters could not avoid these numbers, though they had to be treated with care. In the 1550s John Foxe was part of an influential group of Marian exiles such as John Knox and John Bale, all of whom were concerned with understanding the meaning of Revelation in light of their present situation. The first edition of Foxe's famous *Acts and Monuments of these latter and perilous days*, which depicted the terrible persecution of true saints and martyrs by the Catholic church, appeared in 1563, and there were three further editions. At the end of his life he developed a keen interest in the 'mystical numbers' of prophecy, which first required a grasp of the links between prophetic figures and second demanded that the exegete investigate how these corresponded to historical dates and events. Many of these speculations appeared in a more systematic form soon after his death in his *Eicasmi seu Meditationes in sacram Apocalypsin* of 1587. Foxe's Eureka moment had come, he said, when he had realized that prophetic days and months should not be understood literally, arguing that each prophetic month should be understood as seven actual years. Those working in the English apocalyptic tradition after Foxe would pay particular attention to the seals, trumpets, and vials. Going back to the times of the Apostles, the seals denoted the complete history of Roman persecutions of godly Christians while the sounding of trumpets depicted the rise of Roman Catholicism and then, later on, the rise of Islam. By the early seventeenth century, most interpreters understood the vials to be a description of the downfall of popery as a result of the Reformation.[15]

The apocalyptic studies of the Scot John Napier represented the first major effort to bring the authority of mathematical demonstrations to the understanding of the meaning of apocalyptic images. Napier's *Plaine Discovery of the whole Revelation of St. John* appeared in 1593, motivated in part by the threat posed by the Armada. Napier, who later became famous for his pioneering work on logarithms, placed arguments and mere assertions (e.g. that the Pope was Antichrist) within a mathematical propositional framework, though he also made use of Ramist tables to divide up the subject matter between sacred and imperial history. He used what were by now standard similarity relations between prophetic periods of time and placed them in specific propositions. He adopted the commonplace exegetical view that prophetic days referred to real years, and argued that all those periods equivalent to the 42 months of Rev. 13:5 (in Dan 7:25, 12:7 and Rev. 11:2–3, 12:6, and 12:14) during which the 10-horned beast was to spout blasphemies, referred to the duration in years of the reign of Antichrist. Uniquely for his time, Napier 'synchronized' the trumpets and the vials, so that each trumpet and vial were held to refer to the same period of time. He claimed that each new trumpet and vial (the first occurring with the opening of the Seventh Seal in 71 AD) was sounded or poured at 245 year intervals, meaning that the 7th or last occurred in 1541. Thereafter five angels would act every 'Jubilee' or 49 years, signalling the downfall of Antichristian popery. Finally, with the advent of the fourth of these angels in 1688, the elect would rule in conjunction with the newly returned Christ.[16]

The most influential body of prophetic work in the seventeenth century was produced by Joseph Mede, a don at Christ's College Cambridge. Mede published his short *Clavis* (or 'Key') to the apocalypse in 1627, though an expanded version with a commentary appeared five years later. Going beyond Foxe and Napier, Mede reduced the order of the apocalyptic images to a series of 'synchronisms' that linked these images in terms of the periods of time they allegedly covered. He divided up sacred history into three eras, of which the middle period, that of the Great Apostasy that began towards the end of the fourth century after the birth of Christ, was the focus of his work. Following standard procedure, he linked together the events depicted by the references to 42 months, with the duration of the Great Apostasy. Aside from the obvious connection between these periods, he claimed that other descriptions of events were sufficiently similar that they could be deemed to be synchronal. Since prophetic days meant real years, the Apostasy lasted 1260 years from some point in the late fourth century when the worst perversions of Roman Catholicism spread most quickly. Although he constantly insisted on his orthodoxy as a member of the Church of England, Mede's work was a tremendous resource for Puritans. Not only did Mede strongly identify with the tenet that the pope was

Antichrist, a view that was opposed by the Archbishop of Canterbury William Laud, but his strong insistence on the reality and indeed the imminence of the millennium was a great solace to Puritans who were being increasingly persecuted by Laud. The gist of Mede's findings would become widely known through almanacs and other popular productions in the 1640s, but his major influence would be on later scholarly exegetes such as Newton.[17]

At the end of his life Mede became beguiled by the spectacular use of numerical techniques made by Francis Potter, rector of Kilmington in Devon from 1626 to his death in 1678. Potter had a wide range of interests and talents, and was an inventor (devoted to constructing a perpetual motion machine) and an accomplished instrument-maker. Despite being closely associated with Samuel Hartlib and the Commonwealth regime, he was considered sufficiently proficient as a mathematical practitioner to be elected to the Royal Society in 1663. Potter's most influential text was his *Interpretation of the Number 666* (Figure 7.1), which was published in English in the politically and religiously charged year of 1642 and later translated into a number of European languages. The book contained an encomium from Mede, whose own *Clavis* was translated into

Figure 7.1 The frontispiece to Francis Potter's *Interpretation of the Number 666* (1642).

English at the same time. Mede pointed out that the basic idea of demonstrating various proofs about the nature of the religion of the Beast (by antitypical opposition to the mystical number 144 and its square root) had been known for some time. Nevertheless, he added, Potter had gone far beyond this and had made a 'wonderfull discovery': he had shown how, properly understood, the number allowed interpreters to grasp 'not only the manner and property of that state, which was to be that Beast, but to designe the City wherein he should reigne, the figure and compasse thereof, the number of Gates, Cardinall titles or Churches, Saint *Peters* Altar, & I know not how many more the like.' Mede had first read the book 'with as much prejudice against numericall speculation as might be', but left off with as much admiration as he had come to it with antipathy.[18]

Potter claimed in his opening remarks that – to the ignorant – the work might appear like a 'an intricate labyrinth of curious & unnecessary speculations'. However, the more discerning should either work harder to acquire the necessary knowledge to understand it, or leave it to other students, 'who will, or may draw a probable argument, that this is the true interpretation'. The content of the book was indeed obscure, he admitted, but the time had come for religious mysteries to be revealed. Potter pointed out that prophecy was replete with references to the number 12, but that the Holy Ghost often made use of square numbers. Key here was the mystical number 144, which by Rev. 21:17 was the length (in cubits) of the wall of the New Jerusalem, whose structure represented the hierarchy of the true church. If the number 12 was significant as the root of 144, then by 'antithesis' the godly should perform a similar root extraction ('count the number of the beast') on the major number associated with the church of Antichrist, which was 666. Since it was inexact, its square root clearly delineated some sort of mystery; if it had not been a mystery then Antichrist could have recognized it and avoided its implications. The root was approximately $25\frac{41}{51}$, or more exactly $25\frac{25}{31}$, though Potter was aware that one could obtain any required degree of precision for the value. He took some care to explain why, given its continual presence in the determination of the root, the 'main number' 25 was more important than 26 – despite the fact that the latter was closer to the actual root.[19]

The number 25 always depicted some deplorable character of the church in Scripture, as in Ezekiel 8:16–17 where 25 sun-worshippers turned their backs on God in the inner sanctum of their temple. Since the number 12 was applied throughout Scripture to central features of the true church, the number 25 must depict key aspects of the pseudo-city. Accordingly, the core of Potter's work was a virtuoso demonstration, based on a wide range of sources, of how the number re-occurred throughout elements of Catholic ritual and society, in particular in aspects of the buildings of Rome and the composition of the Roman church. There had been 25 cardinals (who were the diabolical analogue of the apostles) in the original college at Rome, after which there grew to be 25 parishes or tithes, and the original walls of the spiritual Babylon had 25 gates (Figure 7.2). The list grew: the Catholic creed consisted of 25 articles, the Council of Trent had 25 prelates present at 25 sessions, and the number of its decrees concerning faith and reformation was 25. Some references to the number were 'adventitious' and to be ignored, Potter thought, but the Holy Ghost had repeatedly embedded the number in Catholic culture. The *porta sancta* of St Peter's was opened on a jubilee (or half-jubilee) every 25 years, and there were about 25 altars in its vicinity; if this were insufficiently precise for the sceptical reader, it was more pertinent that Christ's five wounds were engraved at five different places on top of every altar. Potter concluded that the number was an 'affected symbolicall device' amongst papists. For those so inclined, his argument was convincing proof that the identity and nature of the bestial religion had been depicted in every aspect of the Roman Catholic world.[20]

Potter's text had been circulated in manuscript in the mid 1630s and having been sent a copy by the Puritan minister William Twisse, Joseph Mede read it in the winter of 1637/8. He told Samuel Hartlib in January 1638 that Potter built on a 'Mathematical ground', though the force of its demonstrations could best be ascertained by those who knew how to extract square roots. Without Mede's knowledge, Hartlib passed Mede's letter back to Twisse, who relayed it

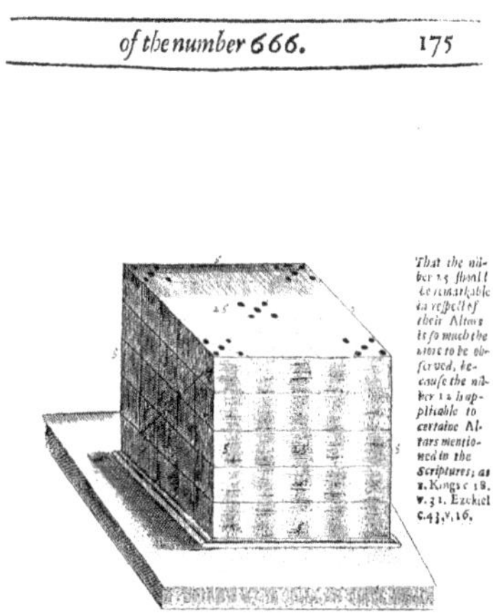

Figure 7.2 Potter's representation of the 25 'characters' (five separate representations of Christ's five wounds) engraved on all Catholic altars.

to Potter – ultimately, much of Mede's letter reappeared as the encomium at the start of the published book. In the spring, Mede reiterated his high opinion of the work to Twisse, saying that if Potter's proof were not true, it still had 'the most considerable Probability' he had met with in a work of its kind. The correspondence between Mede and Twisse reveals fascinating evidence of the extent of interest in apocalyptic investigations within the universities. Twisse told Mede at the end of April that Potter's work had been attacked by Peter Turner, Savilian Professor of Geometry at Oxford, and he passed on to Mede Turner's critique of Potter's work, along with Potter's reply. Turner, who had worked closely with Laud (in the latter's role as university Chancellor) in revising the statutes of the university, and who would be an ardent royalist in the 1640s, had little time for the work. He pointed out that many of the mathematical claims were incorrect, and the supposed reasons offered by Potter were specious – for example, the square root of 666 was not 25. In response, Mede told Twisse that he agreed that with Potter's claim that the root of the number had to be considered as the same type of (non-fractional) integer as the number whose root it was.[21]

Exegetes knew that there was an addictive quality to numerological analysis and they continually warned against overdoing it. On this occasion Mede himself fell victim, and wondered whether there was some special mystery attached to the fractions. He admitted to Twisse that he had been carried away both with the 'remainder' $\frac{41}{51}$, and with Potter's argument (which did not appear in the printed version) that the numerator and denominator were somehow connected to the latitude of Rome. He had gone back to check older books that gave a value for the latitude and longitude of the city, but admitted that he had failed to comprehend Potter's arguments. His own skill in mathematics, he told Twisse, was sufficient to allow him to understand the general case, but not to understand the mathematical details. He had once studied the 'Grounds' of the mathematical sciences but had long neglected them. Otherwise, Mede said he was bowled over by Potter's claim (Figure 7.3) that the similarities between the numbers given in Ezekiel and John for the size of the celestial city could not have arisen by chance – especially because John was unlearned 'and far enough from skill in *Algebraical* subtilties'. This was so extraordinary, he said, that it was the best argument to convince the atheist of the inspired status of the Apocalypse and, indeed, of the entire Bible.[22]

Mede's remarks are also significant for the insight they give into the culture of university apocalypticism. He revealed that he had tried to get his colleagues at Cambridge to read Potter's discourse, but had failed. Twisse had recommended sending it to an unnamed scholar at Cambridge, but Mede responded that

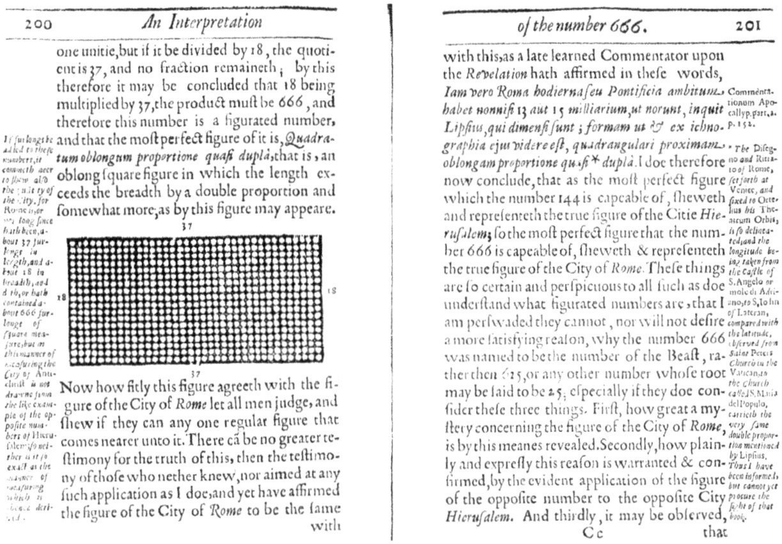
200 *An Interpretation*

one unitie, but if it be divided by 18, the quotient is 37, and no fraction remaineth; by this therefore it may be concluded that 18 being multiplied by 37, the product muſt be 666, and therefore this number is a figurated number, and that the moſt perfect figure of it is, *Quadratum oblongum proportione quaſi duplâ*, that is, an oblong ſquare figure in which the length exceeds the breadth by a double proportion and ſomewhat more, as by this figure may appeare.

37

18 18

37

Now how fitly this figure agreeth with the figure of the City of *Rome* let all men judge, and ſhew if they can any one regular figure that comes nearer unto it. There cã be no greater teſtimony for the truth of this, then the teſtimony of thoſe who nether knew, nor aimed at any ſuch application as I doe, and yet have affirmed the figure of the City of *Rome* to be the ſame with

of the number 666. 201

with this, as a late learned Commentator upon the *Revelation* hath affirmed in theſe words, *Iam vero Roma hodierna ſeu Pontificia ambitum habet nonniſi 13 aut 15 milliarium, ut norunt, inquit Lipſius, qui dimenſi ſunt; formam ut & ex ichnographia ejus videre eſt, quadrangulari proximam, oblongam proportione quaſi * duplâ.* I doe therefore now conclude, that as the moſt perfect figure which the number 144 is capeable of, ſheweth and repreſenteth the true figure of the Citie *Hieruſalem*; ſo the moſt perfect figure that the number 666 is capeable of, ſheweth & repreſenteth the true figure of the City of *Rome*. Theſe things are ſo certain and perſpicuous to all ſuch as doe underſtand what figurated numbers are, that I am perſwaded they cannot, nor will not deſire a more ſatisfying reaſon, why the number 666 was named to be the number of the Beaſt, rather then 625, or any other number whoſe root may be ſaid to be 25; eſpecially if they doe conſider theſe three things. Firſt, how great a myſtery concerning the figure of the City of *Rome*, is by this meanes revealed. Secondly, how plainly and expreſly this reaſon is warranted & confirmed, by the evident application of the figure of the oppoſite number to the oppoſite City *Hieruſalem*. And thirdly, it may be obſerved,

Cc that

Commentationum Apocallyp. part. 2. p. 152.

* The Diſegno and Rittratto of Rome, ſet forth at Venice, and fixed to Ortelius his Theatrum Orbis, is ſo delineated, and the longitude being taken from the Caſtle of S. Angelo or mole di Adriano, to S. Iohn of Lateran, compared with the latitude, obſerved from Saint Peters Church in the Vatican, to the Church called S. Maria del Popolo, carrieth the very ſame double proportion mentioned by Lipſius. Thus I have been informed, but cannot yet procure the ſight of that book.

Figure 7.3 Potter's diagram representing a rectangle with sides equivalent to the divisors of 666, allegedly similar to the shape of seventeenth-century Rome.

although this individual was highly skilled in mathematics, he was so perversely confirmed in his opposition to the notion that the Pope was Antichrist that there was no point in trying to do so. No demonstration could work against someone so imbued with prejudice, he said, but others were also unwilling to read the work – even those who 'delighted in mathematical Speculations' and who did agree that the Pope was Antichrist. Indeed, even though he had a copy of the text divided up into sections and adorned with explanatory figures to make it easier to read, no one so much as turned a page of the work. However, Mede's work did interest the Christ's College student Henry More, whom history has remembered as one of the major proponents of Cambridge Platonism. As I show later, it is more relevant that More was the most prolific published writer on apocalypse in the Restoration, and that he knew Newton at Cambridge, and indeed debated technical exegetical issues with him.[23]

Newton, mathematics, and the Apocalypse

One might assume that Newton rose above the sort of approach adopted by Potter and Mede. Even if he was a devout man, he was also a rational scientific investigator who invoked robust evidence and used well-established techniques to make his case. Yet the mystical numbers in prophecy demanded different styles of reasoning and proof from those applicable in science and mathematics, and Newton made extensive use of standard numerical arguments deployed in the apocalyptic tradition. Indeed, he was just as adept and inventive as was Potter in devising corroborative reasons to support claims regarding the meaning of those numbers. He understood that there was special significance associated with the numbers 2, 7, and 12, though he showed no interest in the Trinitarian 3 or the Potterian 25. Like most others working in the Protestant apocalyptic tradition, he assumed that the Pope was Antichrist and the Man of Sin, designated by the number of the beast. These views were obvious and reasonable to people like Newton, Mede, and More, and like the last two, Newton was not shy of condemning ludicrous views such as the notion that the pope was *not* Antichrist. Like Mede and More, he was wary of incorrect, subjective, or dangerously emotive interpretations of prophetic numbers. As he saw it, there was a well-established tradition of prophetic interpretation that had made great discoveries – most notably in Mede's *Clavis*. Based on Mede's methods, he thought, godly men had a duty in the latter days of the world to use their intellect to understand these figures.[24]

Newton found it harder than he did in the realm of natural philosophy to articulate the epistemic status of his own theological proofs. Early in his career he argued that the only way to secure robust and exact knowledge about the natural world was to ground statements on mathematical principles, though in later years he was content to accept that inductively established general laws (such as that of universal gravitation) were good enough for most purposes. In his apocalyptic treatise, he laid down a series of simple rules that would give the interpreter guidance on how to choose amidst a multiplicity of interpretations, such as making the interpretations as simple and harmonious as possible. Newton drew up his work into a series of 'propositions', which referred back to the rules and a number of definitions. He claimed that these arguments were supposed to remove the uncertainty of private 'fancy', and could offer 'certain' or 'true' interpretations. At one point he even remarked that the reasons he had given should compel the assent of any 'humble & indifferent person' who believed in the truth of Scripture and who read it with sufficient attention. However, he also claimed that he did not believe that they amounted to demonstrations on a par with mathematical proofs. No matter how talented interpreters were, if they lacked the moral and religious prerequisites for the job, they would never understand the meaning of prophecy. Such truths, he said, were not like Euclidean proofs, but were demonstrations to the godly and foolishness to the wicked. Some years later he became irate when Richard Bentley asked him if he could offer a mathematical demonstration that days meant years in prophecy.

Bentley told William Whiston that Newton had refused to do so on the grounds that Bentley was 'invidiously alluding to his being a *mathematician*', and had stated that mathematics had no relevance to the discipline.[25]

In various drafts of an apocalyptic treatise he composed around 1680, he structured his argument in the same way as More and Mede had done in their own works. Identities of both qualitative and quantitative periods of time indicated that they were 'collateral', as did analogies or obvious disanalogies between images. Newton followed nearly all of Mede's synchronisms and in due course would add many more of his own. As for identities, he agreed that the 10-horned dragon in Rev. 12:3 and its historical referent was obviously the same as the 10-horned fourth beast in Daniel 7:7 and the 10-horned beast in Rev. 17:3. Like Mede, he 'placed' the first six trumpets in the sixth seal but unlike Mede, who had located the first six vials in the sixth trumpet, Newton followed Napier (though there is no evidence that he used Napier's work) in synchronizing each successive trumpet and vial so that each one described different aspects of the same event or period. There is no doubt that his downgrading of the significance of the vials, and his relegation of significant prophetic events to the distant past, would have raised serious questions about his orthodoxy if he had let his views be known to others.

As we have seen, central to his scheme was the assertion that the events denoted by each corresponding vial and trumpet were synchronous, a statement he designated as the 'second' proposition. The proof of their synchronicity relied on inspecting Newton's tabulation of the content of each vial and trumpet, which he laid out side by side to show that their descriptions were similar, but not so repetitive as to become 'tautologous'. He was careful to point out that although they referred to the same times and general events, their interpretations depicted different aspects of particular events or periods. The short third proposition dealt with the seven 'thunders' mentioned in Rev. 10:3–4, whose utterances the prophet John was told not to write down. Newton claimed that these 'most probably' denoted the same thing as the seven trumpets and vials, and the fact that the content of their message was not revealed was perhaps because it had been 'sufficiently declared' in the trumpets and the vials. It was not probable, he went on, that the thunders had been mentioned for no reason, and so it was likely that they had been introduced to make up a 'ternary' with the trumpets and vials. The interstices between each thunder, vial, and trumpet thus produced the number of the beast, viz. 666.[26]

In a later draft of this text Newton kept the proof structure of his essay but changed the language from 'propositions' to 'positions'. The identity of the thunders, vials, and trumpets was now asserted as 'Position 2', and he offered a much more extensive analysis in the form of five original arguments to prove that the thunders were coextensive with the trumpets and vials. In one argument he referred to the prophecy of the 'little book' that began at Rev. 10:8, which 'contained' the accounts of the trumpets and the vials. This was different from the prophecy of the thunders, which took up the first seven verses of the 10th chapter. Nevertheless, the little book was mentioned as being in the hand of the angel in Rev. 10:2 and this 'interweaving' of the two prophecies denoted that the events to which they referred were contemporary. This was, Newton explained, on the grounds that 'when two contemporary things are to be described one after another it is ye method of ye holy Ghost to interweave them by mentioning that in ye first place wch is last described, least otherwise they should be taken for successive things'. In referring to the little book, the prophet had gone back 'after ye manner of an Historian' to describe things collateral to what he had treated before. To prove this, Newton offered numerous additional reasons to believe that the prophecy of the thunders and that of the little book were 'collateral'. He concluded that the Holy Ghost had not needed to write down the meaning of the thunders because they were sufficiently described in the trumpets and vials.[27]

In a monumental treatise written in the second half of the 1680s, Newton noted that the true church in the time of the Great Apostasy (in the Seventh Seal) was sometimes represented by the number 2 and sometimes by the number 7. One part of Newton's work addressed the allusions in the apocalypse to the rituals and practices of the Jews, and he wrote at length about the importance of the number 7. Both the Jewish temple, which was the 'scene' of the first few chapters of Revelation, and Jewish legal ceremonies were types of things to come, and the number 7 played a central role in both Jewish and Christian religious practice. Jewish worship was performed every seventh day, and Christians followed suit. The seven trumpets, vials, and thunders alluded to the sacrifices of the seven days of the feast of tabernacles. The seven golden candlesticks, which were seven churches (from Rev. 1:20), were references to the sevenfold candlesticks of the tabernacle, and there were seven stars (the angels of the seven churches of Asia), seven letters sent to those churches (from Rev. 2–3), seven lamps around the heavenly throne (from Rev. 4:5), and a lamb with seven horns and seven eyes (Rev. 5:6). One section of Newton's work (Figure 7.4), devoted entirely to the seven churches, was every bit as scholarly and numerologically inventive as was Potter's performance. The seven churches were actually seven different accounts of the same church in the time of the seventh seal (during the Great Apostasy),

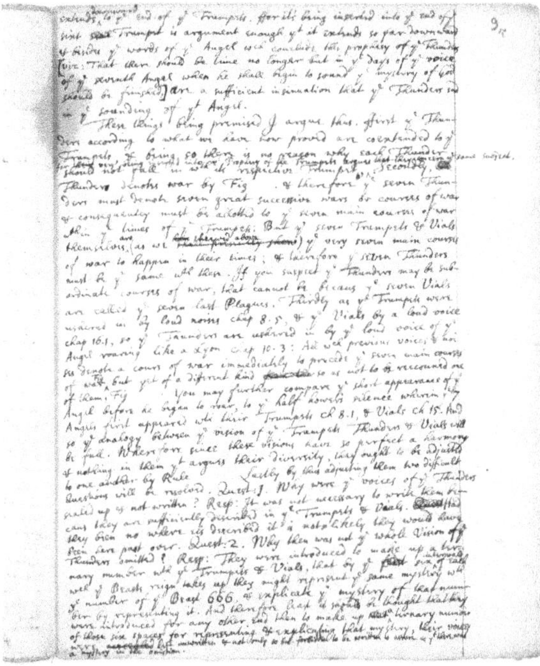

Figure 7.4 Newton's demonstration that the peculiar sevenfold character of the thunders, vials, and trumpets in Revelation could be explained by understanding that the intervals between each item constitute the number of the beast, 666.

while the testimonies of the seven angels were simply repeated references to the Holy Ghost, who spoke to the churches through the letters. These were all references to the true church, and the cognoscenti could determine how to recognize the true nature of Christianity from the same accounts.[28]

Newton's greatest skill rested on his ability to generalize Mede's synchronisms and to show how they corresponded as closely as possible to precise dates or periods of time. Having synchronized most, if not all, of the numbers and images in New Testament prophecy, he collated these with references in Ezekiel, Isaiah, and Daniel. When it came to the historical application of the images, he agreed with virtually all of Mede's dates for the opening of the seals and the pouring of the trumpets (though of course he dissented from him over the pouring of the vials), and followed him in assigning the same dates to the major events. Newton mastered the complex detail of the images and pored over vast swathes of data from church history, interweaving images and historical data to produce interpretations that boasted a degree of precision that far exceeded any of his contemporaries. Where other exegetes made vague allusions to those years where seals might have been opened or trumpets sounded, Newton often adduced evidence showing that key prophetic events had taken place in specific months. In turn, evidence from the past was used to finesse his complex account of how different images cohered with each other. In his approach to mutually fine-tuning theory and data, there are similarities between Newton's *modus operandi* in the *Principia* and in his private exegetical work. It is worth pointing out that he began writing detailed accounts of how prophetic images and numbers had been fulfilled in history many years before he began the *Principia*. However, while his method became the foundation of modern science, his prophetic interpretations remained hidden until the end of the twentieth century.

The mathematical exegete

Despite his best efforts to keep his research secret, Newton's prophetic interests were not entirely unknown. Although he did not use his extraordinary mathematical skills when he interpreted Revelation, his prowess in mathematics attracted the interest of a group of scholars interested in apocalyptic studies during the Popish Plot of 1678–81. During this period, Henry More composed two major works on Revelation and Daniel, and he discussed their contents with Newton. In the second of these works, devoted mainly to those prophecies in Daniel that were referenced in Revelation, More included an appendix in which he attacked some key views held by Newton. Most notably, he condemned as absurdly extravagant the notion that each vial and trumpet referred to the same historical event. More had placed all the vials in the seventh trumpet, and he argued that similarities in their descriptions were insufficiently rich to make each corresponding vial and trumpet 'collateral'. We know that Newton took this text to be an attack on himself, since he responded angrily to More's arguments in annotations in his own presentation copy.[29]

We also know that Newton took umbrage at More's arguments because More said as much in a letter he wrote in August 1680 to John Sharp, a lecturer at St Lawrence Jewry in London and one time student at Christ's. More's letter shows that during a time of great political and religious crisis, a group of people in London were discussing the historical and prophetic significance of the Popish Plot. Sharp and Hezekiah Burton (rector of St George the Martyr in Southwark) had recently put various questions to More about Newton's work on prophecy, presumably because they thought that Newton might have significant things to say about the current situation. More told Sharp that he had been talking to Newton about prophecy for some time, but could not convince him that his views were absurd. In the letter, he admitted that Newton had an astonishing talent for mathematics but countered this by saying that another mathematics professor at Cambridge, probably Isaac Barrow, had told More that he had proved certain prophetic statements 'with Mathematicall evidence'. In his own writings, More repeatedly stressed that his own demonstrations were mathematically certain, and he and his friends must have thought that the Lucasian Professor's mathematical expertise gave him both the skills and authority to understand their own predicament via an interpretation of prophecy. As it turned out, Newton had nothing to say to More about the apocalyptic contexts of late seventeenth-century England, first because to do so might have revealed his own heretical position, and second because he believed that nothing of any great prophetic significance had taken place in the previous three centuries.[30]

Whether he was trying to decode the mathematical laws of nature or the mystical numbers of prophecy, Newton believed that he was a privileged expert – a mathematically adept 'priest' authorized to decipher the mathematical texts used by God. Although he argued that demonstrations in theology and mathematics were at odds with each other, his general approach linked his interpretation of Scripture to the metaphysics that underlay the *Principia*. Like many of his contemporaries, he argued that Moses had 'accommodated' his account of the Creation to the limited reasoning abilities and perceptual faculties of ordinary human beings. There were geocentric passages in Scripture, Newton argued, because God had spoken via Moses to humans in a manner that appealed to what they saw (the sun orbiting the earth) or would have seen had they been present at various moments in Creation. The 'vulgar' experienced only the visible, 'relative' qualities of the world, and could not abstract from their perceptions to understand the real and exact magnitudes of the world that lay beyond. It followed that one could not derive a mathematically precise account of the universe from the Bible, nor, when dealing with the real, absolute world described by the *Principia*, should one deal with 'common' or 'sensible measures'. Only by measurement and mathematical calculation could a sophisticated adept ascertain, by means of his understanding, the true nature of the divine cosmos beyond the veils of experience.[31]

By virtue of the fact that he made little use of his higher order mathematical skills, there is a strange symmetry between Newton's prophetic writings and the approach he adopted in the *Principia Mathematica*. Nevertheless, in both fields, different sorts of mathematical approaches were prominent features of his work. As he saw it, the same God had authored both 'texts', and it required someone with an advanced understanding to decipher each one. Central to Newton's approach was the belief that God was the God of order and precision. He was a stupendously intelligent being, who simultaneously calculated and effected all the infinitely complex motions in the universe. He had also emplotted in the Bible a series of mystical numbers, which learned men like Newton might understand in the last days (which were upon us). Outside of a specific Protestant framework, most people no longer believe, as Newton did, that the quest to understand the meanings of these numbers is a 'rational' exercise. Nor do most scientists accept that the existence of an omniscient and omnipotent deity can be inferred from the astonishing mathematical fabric of the universe. Nevertheless, it was on the basis that God had written both books in a mathematical language that Newton was motivated to test himself, and the limits of human intelligence, against the mind of God.

Notes and references

1. Strictly speaking, the 1706 *Optice* was the first Latin edition of the work, and the 1717/18 edition was the second English edition.
2. For Newton's secretiveness see S. Snobelen, Isaac Newton, heretic: the strategies of a Nicodemite, *British Journal for History of Science*, **32**, 1999, 381–419.
3. See W.R. Albury, Halley's ode on the *Principia* of Newton and the Epicurean revival in England, *Journal of the History of Ideas*, **39**, 1978, 24–43, 27.
4. A.R. Hall and M.B. Hall (eds), *Unpublished Scientific Papers of Isaac Newton*, Cambridge University Press, 1962, pp. 131–3 and 137–8.
5. Hall and Hall, *Unpublished Scientific Papers*, pp. 133–6 and 139–42.
6. 'De motu sphæricorum Corporum in fluidis', CUL Add. Ms. 3965 fol. 47^r. Compare with comments on the inability of humans to grasp all the terms in infinite series expressed in his 1669 work 'De Analysi'; D.T. Whiteside (ed.), *The Mathematical Papers of Isaac Newton*, 8 vols, Cambridge University Press, 1967–81, Vol. 2: pp. 242–3.
7. Newton, *The Principia: Mathematical Principles of Natural Philosophy*, trans. I.B. Cohen and A. Whitman, University of California Press, 1999, pp. 381–2; Whiteside, *Mathematical Papers*, 7: 286–9. More generally see N. Guicciardini, *Isaac Newton on Mathematical Certainty and Method*, Cambridge University Press, 2009, pp. 293–300 (esp. 297 and 300) and 314–5.
8. Newton to Bentley, 10 December 1692, in H.W. Turnbull et al. (eds), *Correspondence of Isaac Newton*, Cambridge University Press, 1959–1977, 3: 233. For the Boyle

lectures, see M. Jacob, *The Newtonians and the English Revolution*, 1689–1720, Ithaca, 1976, 143–200. The sixth, seventh, and eighth lectures printed in Bentley's *Eight Sermons Preached at the Honorable Robert Boyle Lecture in the First Year MDCXCII*, London, 1693, dealt with refuting atheism 'from the origin and frame of the world'.

9. Newton to Bentley, 10 December, 1692; 'Newton Correspondence', 3: 234.
10. Newton to Bentley, 10 December, 1692; 'Newton Correspondence', 3: 234–5.
11. Newton to Bentley, 17 January 1693; 'Newton Correspondence', 3: 239–40. For comments on Newton's reading of Galileo, and the latter's remarks on how the planets could have got their current positions by falling from great heights, see A. Koyré, Newton, Galileo and Plato, in *Newtonian Studies*, Cambridge University Press, 1965, pp. 201–20.
12. Newton to Bentley, 10 December 1692; 'Newton Correspondence', 3: 236.
13. Irenaeus, *Adversus haereses*, bk. 5, ch. 30 §3. For the protestant tradition, see K. Firth, *The Apocalyptic Tradition in Reformation England, 1530–1645*, Oxford University Press, 1978, and for numerology, see V. Hopper, *Medieval Number Symbolism*, Columbia University Press, 1938, and I. Grattan-Guinness, Manifestations of mathematics in and around the Christianities: some examples and issues, *Historia Scientiarum*, **11**, 2001, 48–85.
14. See D. Brady, *The Contribution of British Writers between 1560 and 1830 to the Interpretation of Revelation 13.16–18 (the Number of the Beast): a Study in the History of Exegesis*, Mohr, 1983.
15. Foxe, *Acts and Monuments of these latter and perilous days*, 4th edition, 1583, i, 289–90. See Firth, *Apocalyptic Tradition*, 80–106, esp. 89, 92, 95–6.
16. Napier, *Plaine Discovery of the whole Revelation of St. John*, Edinburgh, 1593. See also Firth, *Apocalyptic Tradition*, 143–49. Prophetic months were calculated as lunar months of 30 days each.
17. See J. Worthington (ed.), *The Works of the Learned and Pious Joseph Mede*, Cambridge, 1672.
18. Mede's text (extracted from his letter to Samuel Hartlib of January 1638) appears at the start of Potter, *An Interpretation of the Number of the Beast*, Oxford, 1642.
19. Potter, *Interpretation*, sigs $2^{r\text{-}v}$, 4^{r}, 5, 13, 34, 43–8, 59, 63–8 (for the square root of 666), 72–7. Potter explained in a letter to William Twisse of 27 June 1638 that the number 144 did not denote the same thing as the numbers of the virgin company or church militant (144,000). The latter was the *anti-numerus* of those who followed the beast, which was not specified in prophecy, though was probably 666,000. See Worthington, *Works*, pp. 856–8.
20. Potter, *Interpretation*, pp. 86, 98–9, 108–14, 125–32, 138, 140–6, 150, 166–71 (for adventitious examples), 176 and 180–90 (Potter's responses to objections that the root was not an integer).
21. Hartlib to Mede, 24 January, 1638, Mede to Hartlib, 29 January 1638, Hartlib to Mede, 6 April 1638, Mede to Hartlib, 16 April 1638, Mede to Twisse, April 1638,

Twisse to Mede, 30 April 1638, Mede to Twisse, 30 May 1638, Mede to Twisse, 23 May 1638; Worthington, *Works*, pp. 877–80 and 851–3.

22. Mede to Twisse, 23 May 1638; Worthington, *Works*, pp. 853–4. Potter told Twisse at the end of June that Mede had misunderstood his argument about the latitude and longitude of Rome, which had been that the kingdom of Antichrist had to fall between the parallels and meridians crossing 41° longitude and 41° latitude, and 51° longitude and 51° latitude; Worthington, *Works*, pp. 856–7. Ironically, although in his intellectual fervour he failed to grasp the completely arbitrary form of the fraction in question, Mede's argument (which he withdrew in July) was more compelling – the current value for the latitude of Rome is 41° 51′.

23. For Restoration apocalypticism, see W. Johnston, *Revelation Restored: The Apocalypse in later Seventeenth-Century England*, Woodbridge, 2011.

24. National Library of Israel, Yahuda Ms. 1.1 fols 4^{r}, 8^{r}. See R. Iliffe, Making a shew: apocalyptic hermeneutics and Christian idolatry in the work of Isaac Newton and Henry More, in J.E. Force and R. Popkin (eds), *The Books of Nature and Scripture: Recent Essays on Natural Philosophy, Theology, and Biblical Criticism in the Netherlands of Spinoza's time and the British Isles of Newton's time*, Kluwer, 1994, 55–88.

25. Yahuda Ms. 1.1 fols 10^{r}, 13^{r}, 14^{r}, 15^{r} and 18^{r}–19^{r}; W. Whiston, *Memoirs of the Life and Writings of Mr. Whiston*, 3 vols, second edition, London, 1753, i. 94. See also R. Delgado-Moreira, Newton's treatise on Revelation: the use of a mathematical discourse, *Historical Research*, **79**, 2006, 224–46. See also Guicciardini, *Newton on Certainty*, esp. pp. 230–3.

26. Yahuda Ms. 1.3, fols 1^{r}, 6^{r}, 8^{r}, 11^{r}.

27. Yahuda Ms. 1.2 fols 6^{r}–9^{r}.

28. Yahuda Ms. 9.2 fols 2–5^{r}, 31^{r}–32^{r}, 42^{r} and 81^{r}–96^{r} (esp. 82^{r} and 84^{r}). For the significance of the number 7, see Hopper, *Medieval Number Symbolism*, pp. 78–88 and more generally see M. Goldish, *Judaism in the Theology of Sir Isaac Newton*, Springer, 1998.

29. Iliffe, Making a shew; More, *Apocalypsis Apocalypseos, (or the Revelation of St. John the Divine Unveiled)* (London, 1680), More, *A Plain and Continued Exposition of the Several Prophecies or Divine Visions of the Prophet Daniel, which have or may concern the People of God, whether Jew or Christian: Whereunto is annexed a Threefold Appendage, Touching Three Main Points, the First, relating to Daniel, the other two, to the Apocalypse*, London, 1681.

30. More to Sharp, 16 August, 1680; M.H. Nicolson and S. Hutton (eds), *The Correspondence of Anne, Viscountess of Conway, Henry More and their Friends, 1642–1684*, Clarendon, 1992, 477–9.

31. Newton, *Principia*, 411 and 413–14.

CHAPTER 8

Maria Gaetana Agnesi, mathematician of God

MASSIMO MAZZOTTI

This chapter explores the religious and mathematical experience of Maria Gaetana Agnesi (1718–99), and aims to show the essential connection between these two dimensions of her life. Agnesi (Figure 8.1) was a mathematician active in the city of Milan and the author of a well-received calculus textbook in 1748, which was translated into French and English.[1] Her scientific achievements were in many ways exceptional – among other things she was the first woman to publish a book of mathematics in her own name, and one of the very few women of her age to be perceived as credible and authoritative by the contemporary mathematical community. I shall argue that there were specific historical conditions that made it possible to establish such credibility, in particular the emergence of a reform movement within Catholicism that fostered mathematical education and the participation of women in social and cultural life. I shall also argue that Agnesi's own commitment to mathematical education was a key aspect of her intense religious experience. It is only through the reconstruction of Agnesi's understanding of God that one can try to understand the meaning of her scientific life.

Figure 8.1 Portrait of Maria Gaetana Agnesi (1718–99), appearing in *Storia delle lettere e delle arti in Italia*, by G. Rovani, Milan 1857.

The lives of Agnesi

Maria Gaetana Agnesi was born in Milan, then the capital of a Duchy under Austrian rule, on 16 May 1718. She was the daughter of Pietro Agnesi (1690–1752), the scion of a family of wealthy merchants who traded in luxury textiles. At the age of 5 she was already known in her native city as a child prodigy, well-versed in languages, memorizing lengthy Latin speeches, and performing effortlessly in front of an audience in her family palazzo. Available descriptions of her skills are to be handled with care, and often contain symbolic elements – for example her alleged ability to speak fluently seven languages – but it is clear that the young girl was highly talented and, most intriguingly for her contemporaries, she would soon excel in the typically masculine art of philosophical disputation. A booklet dated 1727 celebrates Agnesi's wit and the female intellect through a collection of poetry composed within a circle of family friends, and included a Latin oration in defence of the right of women to pursue any kind of knowledge.[2] The oration had been written in Italian by one of Agnesi's tutors, and she had translated and memorized it as part of her studies. In the ensuing years she studied natural philosophy and mathematics with prominent local scholars. Her studies were interrupted in the early 1730s due to a mysterious and persistent malady, in coincidence with a period of repeated performances, the departure of her favourite tutor, and the death of her mother. Her 'convulsions' eluded any diagnosis or treatment until about 1733, when she had apparently recovered and was back to her studies.

Her healing was attributed to the direct intervention of San Gaetano, for whom the family had a particular devotion.

In 1738, at age 20, Agnesi concluded her studies with the publication of her Latin theses, under the title *Propositiones philosophicae* (*Philosophical Propositions*), thus mimicking the experience of male students in contemporary colleges.[3] By this time she had achieved the status of a minor celebrity in northern Italy, and was the protagonist of the *conversazione* that met regularly at palazzo Agnesi. The term '*conversazione*' (literally, 'conversation') was used to refer both to the group of most assiduous frequenters and to the site of the meetings, and was a form of sociability that found expression in many palazzi of the city. A year later, at the height of her career as a *filosofessa* (woman philosopher), Agnesi expressed the desire to abandon the very public life that her father had been imposing on her. She longed for a more secluded life, in which she could dedicate herself entirely to the study of mathematics as well as to charitable activities and devotional practices. She thus asked her father to grant her permission to enter the cloister as an Augustinian nun. A family friend effectively describes Pietro's reaction to such a request: 'it was as if he had been struck by lightning'. Pietro 'did not dissimulate his sorrow at the idea of being abandoned by such a dear daughter, who was, more than the others, the delight of his life'. It's unlikely that he would have been upset by the idea of a female member of the Agnesi taking the veil: two of his sisters and four of his daughters would also do it. Rather, the Agnesi *conversazione* had been entirely constructed around Gaetana's public persona and her performances. At the moment when visibility was crucial to Pietro's strategy of social enhancement, he risked losing what had brought him to the elite's notice in the first place. Not surprisingly, he was eager to discuss and negotiate. Eventually, Agnesi agreed to maintain her lay status but only on certain conditions, which would make her life an unusually private one. Agnesi asked for the freedom to dress simply, thus cutting herself off from her father's conspicuous style. She also asked to be able to visit city churches whenever she wished. Finally she asked to be exempted from attending balls, theatrical spectacles, and other worldly amusements. She had decided to separate herself from the sites and rites of social visibility, and she saw this as a step toward a real and meaningful involvement with the life of her fellow citizens. She also asked to be allowed to volunteer at the Ospedale Maggiore – the ancient city hospital – taking care of poor and infirm women. After an initial resistance, Pietro agreed to all his daughter's requests. In return, she promised to participate occasionally in the conversazioni, although there is no evidence that she ever performed a formal disputation again. Agnesi also promised her father that she would not abandon her studies and publications. However, she wished to give up the study of natural philosophy and deepen her knowledge of mathematics, the science that she favoured above all others.[4]

The following decade of intense mathematical study culminated in the publication of the *Instituzioni analitiche* (1748), a remarkable introduction to the new techniques of differential and integral calculus 'for the Italian youth', and the first book of mathematics to be authored by a woman.[5] The *Instituzioni* were well received in Italy, and were later translated into French and English.[6] In the aftermath of the publication Agnesi was invited to join various literary and scientific academies and, in 1750, she was offered an honorary lectureship of mathematics at the University of Bologna, then under the control of the pontifical government. She did not take up that position though, as she considered her work in mathematics concluded with the *Instituzioni*. Pietro's sudden death in 1752 made it possible for Agnesi to cut the last ties with the world of the *conversazioni*, give up her wealth and inheritance rights, and devote the rest of her life to charitable activities, such as teaching children in parish churches and assisting infirm women at the Ospedale Maggiore. In 1771 the Archbishop of Milan, Giuseppe Pozzobonelli (1696–1783), offered Agnesi the directorship of the female section of the Pio Albergo Trivulzio, a new institution created to house invalid and chronically ill patients from the lower urban social strata. She took up the job with her usual determination, steering the Albergo through the jurisdictional conflicts that characterized the reformist age and the turbulent close of the century. She died of pneumonia at Albergo on 9 January 1799. Milan was under French occupation at the time, and she died a citizen of the Repubblica Cisalpina. All forms of outside ceremony had been prohibited to avoid confrontations between French troops and the local population. Agnesi was buried hurriedly in an unmarked mass grave outside the city walls, together with 15 other women from the Albergo.[7]

Reconstructing the life of Agnesi means to engage with a number of important historiographical issues. These include the relationship between science and religion – her understanding of it, as well as ours – and the way scientific practice, mathematics in particular, was gendered in the eighteenth century. It is precisely the extreme form that the tension between science and religion seems to assume in Agnesi's experience that makes her life especially interesting, and one that can be used to shed new light on the notion of Catholic Enlightenment. Agnesi's contemporaries and her eighteenth-century biographers agreed that her scientific achievements were far from being the only noteworthy trait of her life. Her religious and existential experience was considered equally extraordinary. While the devotions and charitable activities favoured by Agnesi were familiar to the Milanese elites, the intensity of her experience was certainly exceptional. As were her radical life choices: her asceticism, the intensity of her devotional practice, the unpublished mystical-devotional writings, the decision to provide material and spiritual assistance to children and infirm women, the abandonment of the family house, and the renunciation of inheritance. To her contemporary admirers Agnesi's life was exemplary, and it was perceived as a whole, with moral and intellectual virtues – wisdom and piety – sustaining each other. However, by the time of her death, the world of Catholic Enlightenment that had sustained her self-understanding and action had come to an end. One expression of the disintegration of this culture can be found precisely in the transformation of the narratives of Agnesi's life from the 1790s onwards. It is in the revolutionary years that the unity of the early accounts is replaced by the opposite narratives of the heroine of the Catholic reaction versus the daring woman mathematician of the Enlightenment age. Throughout the nineteenth and twentieth century, biographers struggled with what they saw as a fundamental interpretive problem: how to conciliate her devotion with her aspirations of social and cultural change and her admiration for modern science. Typically, the solution was found in giving preeminence to one dimension over the other, hence the spate of nineteenth-century pamphlets on Agnesi as a proto-feminist, an enlightened pedagogue, and a modern mathematician rather than a devout Catholic, a mystic, and a charitable woman. Facing such productions a distracted reader might well think of a case of homonymy. Agnesi's persona was split as neatly as the eponymous protagonist of Italo Calvino's *The Cloven Viscount*, whose good and bad halves had been separated by the sword of a Turkish warrior. If splitting up Agnesi's life has been the most adopted historiographical strategy, others have insisted on the dual nature of her own personality. Thus a fin-de-siècle author defined Agnesi as a 'psychological enigma' – an elegant way to bypass the problem of the tension between tradition and innovation that pervades her life without really engaging with it.[8]

These historiographical approaches have been shaped by precise assumptions about the nature of science and its relationship to religion. Thus, the apologetic lives of the revolutionary period incarnated the concerns and values of the reaction against the alleged conspiracy of the French *philosophes* and the depravation of modern science, while the positivist renderings of the late nineteenth century are a direct expression of the so-called 'warfare thesis' of science and religion. According to this view, the interaction of science and religion engenders always and necessarily some form of open conflict, due to the essential incompatibility of religious dogma and scientific method. The analytical tools shaped by the warfare thesis have thus turned the experience of Agnesi into an enigmatic, and ultimately incomprehensible, historical object. In fact, the very categories of 'science' and 'religion' as referring to two incompatible sets of practices would be meaningless to Agnesi. The conceptual dichotomies that have shaped the narratives of her life should not be taken for granted: they are themselves part of the story that we need to reconstruct and interpret.

The warfare thesis has been long discredited in the history of science, although it seems to enjoy an undiminished popularity in other discourses. Its dismissal has coincided with a profound revision of our understanding of the dynamics of the scientific revolution and of the nature of early modern natural philosophy. The transformation of the scholarship on Isaac Newton (1642–1727) over the last 20 years is emblematic of this broader historiographical re-orientation, which has emphasized the way in which religious concerns and the technical-operative dimension are best understood as integrated and as shaping each other rather than separated and opposed. Newton's religious and alchemical texts have begun to be studied systematically and related to his work in mathematics and natural philosophy only in the late twentieth century (see Rob Iliffe's Chapter 7 in this volume). It is not that these texts were previously unknown, rather that they became truly significant only when the rationalist task of devising an absolute demarcation between science and non-science lost its relevance within mainstream history of science.[9]

Agnesi's unfortunate historiographical destiny is far from being an isolated case, though accounts of her life and work are extreme versions of that attempt to isolate the 'scientific core' from 'the rest' that has plagued the study of most early modern natural philosophers. This work of purification has taken many forms. For example, it has been assumed that after the publication of the *Instituzioni* in 1748 Agnesi must have experienced some kind of religious crisis, which turned her away from the study of mathematics for the rest of her life. One can find this interpretation in most of her nineteenth and twentieth-century biographies. It provides a comforting solution to her 'enigma' by postulating a clear-cut separation between her scientific experience, culminating in the publication of her most famous book, and her religious experience, which would characterize the rest of her life. To reinforce this alleged rupture, some biographers claimed that she did indeed take the sky-blue habit of the Augustinian nuns, a piece of information that found its way into the *Dictionary of Scientific Biography*.[10] In fact, Agnesi never became a nun, and her surviving manuscripts show that intense devotional practices, theological concerns, and social commitment were already important parts of her life during her youth. It is also clear that she did not perceive any radical break as she abandoned the study of mathematics for a full commitment to social work, as she saw her scientific studies and the making of her book as components of her broader apologetic mission.[11]

Entering a conversation

Reconstructing the world of Agnesi and the meaning of her choices is complicated by severe documentary limitations. This is due to various historical contingencies, including the perception that a woman could not be a truly significant protagonist of a philosophical and mathematical debate. And yet, through her surviving manuscripts and the indirect evidence about the management of her family's properties, it is possible to glimpse her world, which for an important part of her life had its centre in the *conversazione* of palazzo Agnesi. While most *conversazioni* were hosted by the great senatorial families, and focused on gambling, the reading of poetry, and informal conversation, what Pietro had built around his talented daughter had a somehow more severe academic form. The soirées at palazzo Agnesi were carefully staged spectacles, typically opened and closed by the virtuoso performances of Gaetana's sister, Maria Teresa (1720–95), who would become a distinguished harpsichord player and composer. Jean-Philippe Rameau was her favourite composer, but she would also play music and arias she had composed herself. Her music and singing framed the different stages of the soirée, providing rhythm and a recognizable theatrical pattern. Visitors included a circle of friends – mostly clergymen and members of the local administrative and scholarly elites – as well as visitors from across Europe. Typically, Agnesi would converse amiably with the newcomers. Then, under the attentive direction of Count Carlo Belloni (1706–47), she would engage in a set of formal debates. Inspired by the academic disputation, these *accademie domestiche* saw Agnesi defending a set of propositions against an opponent who would try to contradict them. The topics were taken largely from natural philosophy, and Agnesi was keen on defending the positions of Isaac Newton and of those contemporary philosophers who criticized late scholastic doctrines. Among her favourite matters were the nature of light and colours, the theory of tides, and the origins of spring waters. Sometimes visitors were invited to step in and debate with Agnesi on topics of their choice, be they philosophical or mathematical. The discussion could thus take less predictable directions, touching metaphysical questions such as the nature of the soul and the relationship between the soul and the body, or the properties of certain geometrical curves. The language of choice would be Latin, so that everyone could understand, although visitors would sometime ask permission to speak in French. At the end of the spectacle, ice cream and sorbet were served by liveried servants.[12]

It is not just the form of the Agnesi *conversazione* that is remarkable, but also its setting. Patrician families in Milan had built their collections around the Renaissance masters, emphasizing the mythological themes that could function as a celebration of the virtues of the house. Ancestral portraits were also particularly relevant in a city like Milan, where local politics had depended on connections with distant capitals like Madrid and Vienna. Signs of political affiliation and patronage were proudly exhibited through the clothing and decorations of portrayed family members. One would look in vain for such paintings in the rooms of the palazzo Agnesi. Rather, the two sisters performed against a severe background of religious paintings in which the representation of the passion, death, and resurrection of Christ had a pre-eminent role. The display also included a number of landscapes and marinas. The taste that had shaped the collection was similar to that of high-ranking Milanese ecclesiastics, who were traditionally among the foremost art collectors in the city. Thus the collection of Archbishop Pozzobonelli was characterized by the same thematic dichotomy between sacred subjects and landscapes, an expression of the moral classicism of early eighteenth-century Arcadian culture. Another common trait was the complete absence of minor genres, such as pulcinellas, beggars, and macabre scenes, that were favoured by the local aristocracy.[13]

One could not help but notice that it was Count Belloni, a family friend, who played the host to this *conversazione*. While carefully directing events from behind the scenes, Pietro always maintained a low profile, while there is no evidence that his wives – he married three times – ever played a significant role in the *conversazione*. Belloni, from a family of the provincial aristocracy, aimed at entering the Milanese stage and had ambitions that could not be fulfilled in the provinces. To Pietro, he offered his easy speech and the savoir-faire of a well-groomed and connected patrician. Pietro needed both of these, as he was a new man, the first in his family to move away from trade and warehouses, embracing the lifestyle of the aristocracy. But Pietro had no titles, and those that were attached to properties he bought were not something that would let him join the patrician elite. Thus his sons could not enter the colleges for the nobility, nor could he access those administrative positions that were reserved for members of the senatorial families. Pietro's entire life became devoted to a strategy of social enhancement that eventually proved to be too expensive to be sustained, and caused the family's economic ruin. When the coat of arms of the Agnesi was recorded in the Heraldic Tribunal of the state of Milan in 1773, Pietro was long dead and the family had lost most of its properties and wealth.[14]

Understanding this family strategy is essential to contextualize the nurturing and showcasing of Agnesi's talent, and to interpret the distinctive features of the Agnesi *conversazione*, including its distance from contemporary aristocratic models. It is not just that this *conversazione* was different as a form of sociability: the kind of knowledge that was being produced, discussed, and promoted within it was also different. Natural philosophy and mathematics as understood within the *conversazione* and as practised by Agnesi, for example, were markedly different from the hegemonic Jesuit model. It is not a coincidence that Jesuit science was absent from the otherwise well-stocked Agnesi library, and that the *conversazione* did not include any of those prestigious Jesuit mathematicians who lived in the city at the time (such as Girolamo Saccheri and Tommaso Ceva), and were organic to the great aristocracy. The cultural and political referents of the Agnesi were to be found elsewhere, along networks that connect the family to Vienna, on the one end, and Rome, on the other. In an effort to bypass the sclerotic social structure of his native city, Pietro looked indeed for distant but powerful patrons. One was the emperor himself, keen on rewarding faithful new men from the provinces of the empire: Pietro will indeed receive an imperial fief in 1740. The other was the Archbishop of Milan and, through him, that part of the Roman Curia that aimed for a renewal of Catholic culture, an engagement with modern philosophy that would go beyond the Jesuit positions, and the return to a more sober and pure religiosity.

I refer to this reform movement as 'Catholic Enlightenment', a term used by historians such as Emile Appolis to describe a 'third party' between the conservative front and the varied world of Jansenism, which was active in the early to mid eighteenth century.[15] It is in this specific and rather narrow meaning that I describe Agnesi's world as the world of the Catholic Enlightenment. The movement found some of its key texts in the writings of Lodovico Antonio

Muratori (1672–1750), and achieved maximum visibility in the 1740s, when it was backed by learned cardinals, literary periodicals, and the pontiff Benedict XIV, who considered Muratori 'the light of Italian science'.[16] Historians have explored the historiographical, liturgical, and theological dimensions of this movement, and its aspiration to return to the experience of an idealized primitive church – hence the relentless critique of baroque devotions and modern, especially Jesuit, theology. But the battle against the perceived rampage of impiety and heresy needed more than austere religiosity. Enlightened Catholics believed that religion was not an obstacle to but rather the means of transforming society, and that this transformation should necessarily pass through a critique of the traditional ways in which knowledge was produced and legitimated. The transformation of the education system envisaged by these reformers can be described succinctly as a dismantling of the Jesuit curriculum. In particular, scholastic metaphysics had to be expelled from the study of natural philosophy, and replaced by the study of Descartes, Malebranche, Newton, and the Dutch experimentalists. The Catholic Church, they argued, could respond effectively to the challenges of the new century only through this alliance of primitive theology and modern science. It is emblematic that the offer of an honorary lectureship of mathematics made to Agnesi by the University of Bologna (1750) was propitiated by the pontiff himself. While the theology and liturgical reformism of the Catholic Enlightenment have been studied in detail, the role of this movement in shaping contemporary scientific life has attracted little attention, and the very notion of 'Catholic Enlightenment' is hardly familiar to historians of science. Agnesi's existential and scientific experience lies precisely at the intersection of the Catholic Enlightenment and the contemporary practices of natural philosophy and mathematics.

The enlightenment of Agnesi

The *Propositiones philosophicae* (1738) offer a good insight into the kind of natural philosophy in which Agnesi was trained, and some indications of her own distinctive take on important philosophical issues. These propositions are a fascinating medley of scholastic, Cartesian, and Newtonian ideas, which was indeed a trait of much early eighteenth-century scientific culture in continental Europe. The organization of the materials seems to follow a rather traditional structure: logic, ontology, pneumatology, general and particular physics – not unlike a course of philosophy in a contemporary Jesuit college. Even the main textbook she used for her studies – Edmond Pourchot's *Institutiones philosophicae* (1695) – would have been familiar to college students. If one goes beyond the formal structure of her training, however, and looks at her manuscript notes as well, the differences begin to emerge. For example, while college students would focus on logic, metaphysics, and general physics, Agnesi moved quickly through them, working more extensively on particular physics and mathematics instead. Logic and metaphysics are simply the occasion to reflect on methodological and epistemological issues that she saw as preliminary to her scientific work. This was a complete reversal of priorities with respect to traditional college education. In other words, the materials used are largely the same, but they are used differently. The *Propositiones* also give us a good sense of the modern authors that Agnesi managed to integrate in her curriculum. The epistemological framework is Cartesian, as is the emphasis on evidence and the discussion of the conditions of truth of an idea. It is, however, a decidedly apologetic Cartesianism, filtered through the lens of a devout sensibility. A belief in the fundamental harmony of faith and reason shapes Agnesi's understanding of the process of cognition, and makes her turn to authors like Nicholas Malebranche (1638–1715), probably the single most influential author in her studies. Agnesi is not interested in Malebranche's problematic metaphysics and theology, but rather in his effort to integrate modern science and the Catholic tradition. Thus Agnesi is interested in the doctrine of occasionalism – according to which all events are caused directly by God – and the image of knowledge as the product of the 'vision in God'. The belief that human reason is an essentially passive faculty, which requires divine illumination in order to 'perceive' its objects, will return often in Agnesi's own notes. The framework of general physics is also Cartesian, although this does not prevent Agnesi from enthusiastically embracing many Newtonian 'doctrines', from mechanics to the origin of celestial bodies, to the theory of colours. There seems to be an aesthetic appreciation of Newtonian natural philosophy, which she defines a 'most beautiful and simple theory'.[17] Already at this early stage it is possible to see that Agnesi is especially fond of mathematics, as a source of certainty and of unparalleled intellectual pleasure.

After the renegotiation of her lifestyle at the end of 1739, Agnesi gave up the study of natural philosophy to concentrate on the study of mathematics and the making of her main book. This was in many ways a mystifying choice. While topics in natural philosophy could make for sparkling conversation, it was much less clear that this could happen with mathematics. In fact, Francesco Algarotti's bestselling *Newtonianism for Ladies* (1737) had explicitly banned mathematics from genteel conversation.[18] Not surprisingly, her tutors tried to dissuade her from this choice, but to no avail. One should not think, however, that after 1739 Agnesi had disengaged completely from the world – quite the contrary. She intensified her charitable activities, mainly through her participation in some of the numerous congregations that articulated religious life in Milan. She volunteered at the Ospedale Maggiore, and she set up a small infirmary in a corner of the palazzo where, much to the dismay of her father, she would welcome poor and infirm women. Agnesi would also teach children to read and count at Sunday schools in her neighbourhood. On each school day, she would 'unfailingly' be among the children, in a corner of the church where benches could be arranged in a circle. Standing at the centre, she would use the figures and words from the catechism to teach them how to read, as well as big boards with letters on them. Pupils ranged in age from 5 to 14, and were divided into three progressive classes; occasionally, illiterate women would also join in. At the same time, Agnesi was intensifying her theological study and devotional practices, revealing a clear inclination for the ascetic spirituality that characterized Catholic religious reformers.

Profoundly interested in mathematics but still unclear about the nature of her possible contributions, Agnesi began by planning a commentary of Guillame de L'Hospital's (1661–1704) treatise on curves, to make it more accessible to students. Gradually, however, she came to believe she could work on a much more ambitious project: an introduction to calculus that would guide the beginner from the rudiments of algebra to the new differential and integral techniques. This would be a great work of synthesis, aiming at a clear presentation of materials that were mostly written for specialists, in Latin, French, or German, and published in hard-to-find journals. During the making of her book Agnesi interacted with leading Italian experts, such as Jacopo Riccati (1676–1754). Her correspondence with Riccati sheds light on some of the distinctive traits of Agnesi's mathematical style.[19] Thus we learn that she was consciously keeping out of her book anything that had to do with empirical states of affairs and with possible applications. This choice was at odds with the practice of most contemporary specialists, including Riccati, whose research was guided primarily by the physical meaning of formulas. Also in marked disagreement with leading practitioners was Agnesi's geometrical style, originated from her essentially geometrical understanding of algebra and calculus. This explains, among other things, her interest in Newton's fluxions, and the ease with which her work was translated into English for a British audience.[20] At a time when the practice of calculus on the continent was moving away from its immediate geometrical meaning, Agnesi was aiming at rediscovering those techniques of Cartesian geometry designed to bridge the gap between the two fields. The single most striking feature of the book, however, is its lack of examples of applications of calculus to mechanics or experimental physics. Agnesi claimed that she wanted to limit her book to 'pure analysis' and to its geometrical applications. Thus she left out anything that had to do with physics in order to preserve the simplicity, rigour, and evidence that she believed were proper to Greek geometry. Indeed the geometrical figure that is commonly associated with Agnesi's name is a curve characterized by interesting metrical properties, but that had no physical significance at the time. In the English-speaking world this curve is still known as the 'Witch of Agnesi'. John Colson, the Lucasian Professor of Mathematics who translated the *Instituzioni* in 1801, interestingly confused the Italian name for this curve (*versiera*) with that for 'she-devil' (*avversiera*) (Figure 8.2).

Agnesi's inclination for pure mathematics explains her interest in the Oratorian mathematical tradition, and especially authors like Malebranche and Charles Reyneau (1656–1728). The reference to Reyneau as her main source of

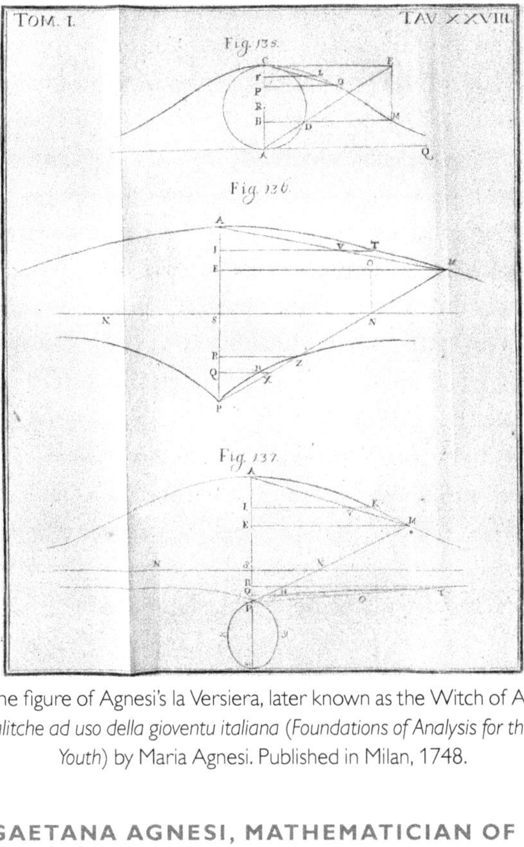

Figure 8.2 The figure of Agnesi's la Versiera, later known as the Witch of Agnesi. Given in *Instituzioni analitche ad uso della gioventu italiana (Foundations of Analysis for the Use of Italian Youth)* by Maria Agnesi. Published in Milan, 1748.

inspiration for the *Instituzioni* has surprised many readers, as he was hardly considered a valuable resource for mid-eighteenth-century continental mathematicians, and was usually described as a pedantic and obscure writer. To Agnesi, however, Reyneau offered an example of how to understand recent development in calculus within an essentially Cartesian framework. Furthermore, he too was uninterested in the empirical applications of calculus, and invested the practice of mathematics with important spiritual meaning. Far from being an odd curiosity, Agnesi's interest in the Oratorian tradition gives us a lead to understand her motivations and her otherwise puzzling choices.

Why did Agnesi decide to write a calculus textbook when there was little or no interest in Milan for these mathematical techniques, and they were not being taught in any local school or university? And why did she write a textbook that focused on the foundational and purely geometrical dimension of calculus, ignoring those applications that made it interesting in the eyes of most contemporary specialists? I believe that these questions can only be answered by considering the pre-eminent role of intellectual exercise within Agnesi's personal religious experience. Among her surviving manuscripts is an undated text titled *The Mystic Heaven*, which is especially significant in this respect.[21] This text is distant from the visionary mystical tradition of the baroque age. Rather, it describes a contemplative practice that moved from meditation on the mysteries of the passion and led to the gradual elevation of the soul towards God. The style and themes are largely derived from a tradition that exalted personal love for Christ and the contemplation of his death and resurrection – one that held a special position in the spirituality of the Theatine congregation. Key to this practice was the ability of the believer to move from contemplation of the concrete objects of the passion, or the pictorial representation, up to contemplation of profound truths of faith and, ultimately, to the contemplation of the Holy Trinity. Agnesi described this process of elevation as built upon the use of two faculties: intellect and will. It is only their cooperation (*cospirazione*, she said) that allows the believer to experience the final vision of God. In other words, Agnesi saw no conflict between rational knowledge, contemplation, and mystical experience. A clear and robust intellect is indeed necessary to support a spiritual experience such as the one sketched above.[22]

There is one notion that nicely captures the intersection of intellectual ability and spiritual achievement as understood by Agnesi: the 'capacity of attention'. Many early modern devout authors – most notably Malebranche – used the term 'attention' to describe a particular state of mind, a form of intense concentration that was seen as a necessary condition for contemplative practices, and more generally for a fulfilling spiritual life. Malebranche would refer to attention as a 'natural prayer'. At the same time, this state of mind is described as a prerequisite for the investigation of nature, and in particular for the study of mathematics. Wary of baroque piety and what Muratori called 'disorderly devotions', Agnesi highly valued the capacity of concentration and a well-trained intellect. Superstitious devotions stimulated and relied upon fantasy and imagination, inducing a credulous and fatalistic attitude rather than a clear understanding of one's religious duties. Believers should rather train their intellect and ground their spiritual experience upon it. And which is the best way to exercise the intellect? The study of mathematics, geometry in particular. Agnesi believed that calculus was the subtlest branch of mathematics, and therefore the one that required the highest level of concentration and the strongest intellect. Once one realizes that for Agnesi the pre-eminent role of calculus is precisely that of exercising the intellect, it becomes understandable why she decided to completely ignore its empirical and applicative dimension – a move that puzzled many contemporaries, as well as later historians of science.[23] This notion of 'attention' finds a striking visual representation in the paintings of Giuseppe Petrini (*c.* 1677–1755; Figure 8.3), whose anti-baroque style aimed to celebrate and promote a religiosity built upon the values of the Catholic Enlightenment. Petrini had a keen interest in representing states of religious and intellectual absorption. His saints, evangelists, astronomers, and philosophers (Figure 8.4) are captured in that particular state of attention that was a prerequisite for the acquisition of both true knowledge and divine enlightenment.[24]

Figure 8.3 *Education of the Virgin*, by Giovanni Antonio Petrini (1744). Courtesy of the Parish of San Antonio Abate, Lugano. Photo by Roberto Pellegrini.

Figure 8.4 Giuseppe Antonio Petrini, *Filosofo* (c. 1735). From *Giuseppe Antonio Petrini*, ed. Rudy Chiappini (Milan: Electa, 1991), 161.

What kind of God was Agnesi's God? Her anti-baroque stance and her rejection of what she considered modern theological aberrations – first of all Jesuit moral theology – made her turn to the fathers of the Church, Augustine in particular. Like her fellow enlightened Catholics she strived for a return to the alleged pure religiosity of the primitive church, shedding the historical stratifications that had polluted that original theological and liturgical simplicity. This was not an optimistic, benign theology. The God of these Catholic reformers is distant, enigmatic. It is a God far removed from human affairs. Hence, in spite of her fervent Newtonianism, Agnesi is obviously uninterested in pursuing any kind of natural theology. While, as I have tried to show in the previous sections, there is a profound apologetic dimension to her scientific work, this does not take the familiar form of, say, the argument from design. Science and mathematics are important components of a Christian upbringing, but not because they reveal God's creative action in the world. In fact, Agnesi is always very careful to avoid constructing any argument of this kind. She probably considered this inductive process simplistic, if not blasphemous. At the root of her contempt for natural theology was the boundary between physics and metaphysics, which she perceived as sharp and unbridgeable. It was a veritable hiatus, which made it futile and dangerous to produce a discourse on God based on the experience of the natural world. Devout scholars should not try to gaze into the mind of God through the marvels of creation. Rather, they should discipline their intellect through strenuous exercise, and in this way try to achieve a proximity to God that could only be the result of intense meditation and ascetic practices. The devout scholar cannot second-guess God and penetrate God's thoughts, but they can elevate themselves to an experience of God through an integration of intellect and what Agnesi at times called 'burning love'.

The world of the *filosofesse*

Agnesi was not the only *filosofessa* active in northern Italy in the first half of the eighteenth century. A handful of women gained visibility as competent scholars in that period, the most famous and successful being certainly Laura Bassi (1711–78), who graduated at the University of Bologna in 1732 and taught experimental physics there for much of her life.[25] In her pioneering study of the gendering of modern science, Londa Schiebinger detected the unusual presence of learned women-philosophers in northern Italy during the first half of the eighteenth century. She described it as an interesting anomaly in search of an explanation: 'Italy was an exception in Europe, and little is known about why women professors were acceptable to the Church and university.'[26] We now know much more about the world of the *filosofesse*, and the way they could achieve visibility and legitimation though specific networks of patronage.[27] The reconstruction of Agnesi's experience can contribute to shed light on this anomaly by bringing to the surface its relation to the culture of the Catholic Enlightenment. It was indeed this reformist culture that offered to Agnesi a set of resources and possibilities that were absent in other European settings. For one thing, enlightened Catholics placed particular significance on the contribution of women to religious and social life. They also celebrated a new kind of lay sainthood that engaged with the daily problems of the poor, and valued popular education and the role of the intellect in spiritual life. Most importantly, enlightened Catholics were not inclined to believe that the mind was gendered, as sustained by certain recent scientific doctrines, and that the female mind was unfit for scientific and especially mathematical studies. It is indeed during the first half of the eighteenth century that a new kind of misogynist literature gains strength, which emphasized the physical basis for women's intellectual inferiority.[28] While traditional arguments against women's education and participation in intellectual life were based on the moral and social disruption that would follow from them, these new philosophical arguments provided an alleged objective assessment of the inadequate physical structure of the female body and mind, shifting the debate progressively toward the terrain of physiology. It was argued, for example, that the woman's body is less vigorous than the man's, and that physical vigour depends ultimately on the 'consistency' and 'elasticity' of the fibres that constitute the organism. Because of its reproductive function, the female body is richer in fluid and as a consequence its fibres contain more liquid, which makes them less elastic and solid. Anatomical observations were cited to confirm that there are important differences in the constitution of the female body. Anatomy, however, was less useful for inspecting the brain, so here arguments shifted to behavioural observation and analogy. Many enlightened authors shared the belief that ideas were always accompanied by some movement of the brain. More precisely, the vibrations of the tiny extremities of the nervous filaments were associated with intellectual activity: the wider and more prolonged the vibration, the more vigorous and general the idea produced. In this way the notion of attention was translated in mechanistic terms: the greater the number of vibrations and their duration, the more the spirit can concentrate on one idea, examine it, analyse it, and compare its parts. It followed that male cerebral fibres, being drier, are capable of longer and more ample vibrations, from which complex and abstract ideas originate. Mathematics and fluxional calculus are thus typical examples of intellectual activities that are barred to women because of their physical constitution. The new physiological reinterpretations of traditional beliefs about female weaknesses and inadequacies would become predominant in the scientific discourse of the mid and late eighteenth century. By contrast, the defence of an essentially Cartesian and anti-materialist image of the mind was at the core of the explicit defences of the right of women to study the 'sublime sciences' that were produced within the culture of the Catholic Enlightenment.[29] Agnesi participated directly in this debate. She rejected the three main kinds of objection to women's study: tradition, the social disruption that would follow from this concession, and the alleged inability of the female mind. She argued that learned women would be better daughters and wives because they would be more aware of their religious and social duties, and that there were plenty of historical examples that run against the argument of tradition. As for the nature of female mind, she believed that a rigid distinction should be maintained between someone's intellectual ability and their bodily structure, as the activities of the human spirit are not influenced in any way by the material dimension. Interestingly, she referred to the opinion that the weakness of the female mind is based on the weakness of the female body as a recent philosophical aberration.[30] A few women within the Catholic Enlightenment movement read these and other similar claims as the long-awaited end of the 'tyranny' of men in education and science.[31] Most sympathizers, however, opted for a moderate view, according to which a few extremely talented women could be allowed to enter traditionally masculine social spaces such as scientific academies and universities, but indeed only as exceptions. The question of female education was not perceived as equally relevant by all enlightened Catholics – Muratori himself was ambiguous on this point.[32] In any case, the Catholic Enlightenment created the conditions for the emergence of an alternative discourse on the female mind, and fostered various initiatives in women's education. The archdiocese of Milan, for example, proved to be a veritable laboratory where female religious orders such as the Ursulines, as well as lay groups, experimented with new forms of female education. The local Church authorities supported these initiatives, as they were believed to favour the promotion of a new generation of learned and virtuous women.

There were various reasons why enlightened Catholics tended to emphasize the role of women in a Christian society. For one thing, women were one of the social groups that the Church was now willing to mobilize within its institutional network – especially women from the merchant and professional classes. This was supposed to balance the growing detachment of the traditional elites from the collective rites of religious life. Artisans and 'low people' were specifically targeted by a renewed missionary initiative, and new spaces were opened for members of these groups to participate actively in the life of their parishes. The Church was looking for charismatic figures, both men and women, often from lowly backgrounds, who could promote the new ideals of civic devotion and lay sainthood. Centred on social utility, these ideals contrasted bluntly with the baroque models of mystical and visionary sainthood. Education was thus considered necessary for young women to shape their devotion and prepare their future participation in the religious life of the community. Their role as mothers and first educators of their children made such education all the more relevant. It should be noted that, in the same period, Pope Benedict XIV modified canon law so that women as well as men could produce evidence during processes of beatification and canonization, thus providing them with unprecedented social legitimacy. This new emphasis on female education found an interesting manifestation in contemporary visual culture: during the early eighteenth century, paintings representing the education of the Virgin were repeatedly commissioned in northern Italy, particularly by teaching orders. These compositions portrayed the Virgin as a young girl who follows the lines of the Holy Scripture with her finger as she is being taught to read by her mother. Her father is generally present but he looks distracted, if not drowsy. In 1744 the anti-baroque painter Petrini created an altarpiece on this subject for a Somaschan college in Lugano. Like the rest of Petrini's productions, it was a most effective visual translation of the ideals of the Catholic Enlightenment.

It was in this context that Agnesi and a few other talented women were able to use the network of alliances and resources of the Catholic Enlightenment to establish themselves as credible and legitimate scholars. The phenomenon of the *filosofesse* reached its apogee around 1750, with the invitation to Agnesi to join Bassi at the University of Bologna as a lecturer. That gesture was emblematic of Benedict XIV's strategy of bringing back lustre to the ancient university, and once again placing the Catholic Church at the centre of the European philosophical debate. Far from being a curious anomaly, the *filosofesse* should be seen as an important component of the world of the Catholic Enlightenment. Their experiences effectively illustrate some of the distinctive traits of this movement, and its otherness with respect to other strains of Catholic and enlightened culture. Agnesi's religious experience and her direct engagement with social reform profoundly shaped her mathematical experience, and invested it with meaning. Searching for Agnesi's God, therefore, is not a marginal task, but the only way we have to try to hear her long-lost voice.

Notes and references

1. M.G. Agnesi, *Instituzioni analitiche ad uso della gioventù italiana*, 2 vols, Nella Regia-Ducal Corte, Milan, 1748.
2. M.G. Agnesi, *Oratio qua ostenditur: Artium liberalium studia a femineo sexu neutiquam abhorrere*, Malatesta, Milan, 1727.
3. M.G. Agnesi, *Propositiones philosophicae*, Malatesta, Milan, 1738.
4. A. Frisi, *Elogio storico di Maria Gaetana Agnesi*, Galeazzi, Milan, 1799, pp. 28–30.
5. Agnesi, *Instituzioni analitiche*.
6. M.G. Agnesi, *Traités élémentaires de calcul différentiel et de calcul integral*, Jombert, Paris, 1775; M.G. Agnesi, *Analytical Institutions*, Taylor & Wilks, 1801.
7. Frisi, *Elogio storico di Maria Gaetana Agnesi*, pp. 104–8.

8. M. Mazzotti, *The World of Maria Gaetana Agnesi, Mathematician of God*, Johns Hopkins University Press, 2007, pp. xi–xii.
9. J. Force and S. Hutton, *Newton and Newtonianism: New Studies*, Kluwer, 2004. M. Mazzotti, The two Newtons and beyond, *British Journal for the History of Science*, **40**, 2005, 105–11.
10. E. Kramer, 'Maria Gaetana Agnesi' in C.C. Gillispie (ed.), *Dictionary of Scientific Biography*, Vol. 1, pp. 75–7.
11. Mazzotti, *The World of Maria Gaetana Agnesi*, pp. 145–6.
12. C. de Brosses, *Lettres d'Italie du Président de Brosses*, 2 vols, Mercure de France, Paris, 1986, pp. 135–52; 162–4.
13. Mazzotti, *The World of Maria Gaetana Agnesi*, pp. 8–9.
14. Mazzotti, *The World of Maria Gaetana Agnesi*, pp. 9–16.
15. E. Appolis, *Entre jansénistes et zelanti: le 'tiers parti' catholique au XVIII siècle*, Picard, Paris, 1960; B. Plongeron, Recherches sur l'Aufklärung catholique en Europe occidentale, 1770–1830, *Revue d'histoire moderne et contemporaine*, **16**, 1969, 555–605; L. Rogier, L'Aufklärung cattolica, *Nuova storia della Chiesa*, **4**, 1975, 151–74; M. Rosa, *Cattolicesimo e lumi nel settecento italiano*, Herder, Rome, 1981.
16. L. von Pastor, *Storia dei papi*, Desclée, 1955, Vol. 16, p. 148.
17. Agnesi, *Propositiones philosophicae*, p. 81.
18. F. Algarotti, *Il newtonianesmo per le dame, ovvero dialoghi sopra la luce e i colori*, Naples [but Milan], 1737
19. M.L. Soppelsa, Jacopo Riccati—Maria Gaetana Agnesi: carteggio 1745–51, *Annali dell'Istituto e Museo di Storia della Scienza di Firenze*, **10**, 1985, 117–59.
20. Agnesi, *Analytical Institutions*.
21. M.G. Agnesi, Il Cielo Mistico, cioè contemplazione delle virtù, de' Misteri e delle Eccellenze di Nostro Signore Gesù Cristo [n.d.], in M.L. Anzoletti, *Maria Gaetana Agnesi*, Cogliati, Milan, 1901, pp. 435–72.
22. Mazzotti, *The World of Maria Gaetana Agnesi*, pp. 90–2.
23. C. Truesdell, Maria Gaetana Agnesi, *Archive for the History of the Exact Sciences*, **40**, 1989, 113–42.
24. Mazzotti, *The World of Maria Gaetana Agnesi*, pp. 85–7.
25. P. Findlen, Science as a career in Enlightenment Italy, *Isis*, **84**, 1993, 441–69; M. Cavazza, Laura Bassi e il suo gabinetto di fisica sperimentale: realtà e mito, *Nuncius*, **10**, 1995, 715–53.
26. L. Schiebinger, *The Mind has No Sex: Women in the Origins of Modern Science*, Harvard University Press, 1989, p. 16.
27. P. Findlen and R. Messbarger, *The Contest for Knowledge: Debates over Women's Learning in Eighteenth-Century Italy*, Chicago University Press, 2005.

28. A. Conti, *Prose e poesie*, Vol 2, Pasquali, Venice, 1756, pp. lxv–lxxv; M. Cavazza, Women's dialectics, or the thinking uterus: an eighteenth-century controversy on gender and education, in Lorraine Daston and Gianna Pomata (eds), *The Faces of Nature in Enlightenment Europe*, BWV, 2003, pp. 237–57.
29. G. Bandiera, *Trattato degli studi delle donne*, 2 vols. Pitteri, Venice, 1740.
30. Agnesi, *Oratio qua ostenditur*.
31. Agnesi, *Oratio qua ostenditur*, p. 68.
32. Mazzotti, *The World of Maria Gaetana Agnesi*, p. 142.

CHAPTER 9

Capital G for Geometry: Masonic lore and the history of geometry

SNEZANA LAWRENCE

Thus grew the tale of Wonderland:
Thus slowly, one by one,
Its quaint events were hammered out -
And now the tale is done,
And home we steer, a merry crew,
Beneath the setting sun . . .

(**Lewis Carroll**, *All in the Golden Afternoon, 1865*)

We meet as masons
free and true, and when
our work is done,
the merry song and
social glass is not
unduly won.
And only at our farewell
pledge is pleasure
mixed with pain,
happy to meet, sorry to
part, happy to meet
again . . .

(Masonic farewell toast, c. 1860)

Many generations in the history of humanity have dreamt of unity and brotherhood of some kind. This dream arises more strongly and clearly in the midst of wars, and the English Civil War of the seventeenth century was no exception.[1] Amidst the destruction and cruelty, a new type of social organization was hence imagined by men from both sides of the conflict, built on the belief that to build a land of peace and prosperity men must adhere to certain principles of conduct that would stand them in good stead for the duration of their lives.

The peculiarity and longevity of this organization seem to both be linked to a mathematical discipline, and whilst the organization itself does not purport to be 'religious' the various elements of its practices certainly make it spiritual.[2] The organization in question is that of Freemasonry, and the mathematical discipline on which the Masonic system of beliefs is based is geometry.

Whilst other contributions in this book investigate beliefs that mathematicians developed through their work and deal with questions such as 'Does God exist?' and 'Can the existence of the deity be proved mathematically?', this chapter looks at how a popular belief that was to unwittingly become linked to a mathematical technique became synonymous with a certain popular perception of geometry.

Why would Masonic geometry be important to us today, and in this volume? Three reasons can be put forward. First, the ideals of Masonic thought influenced various aspects of intellectual history related to the development and understanding of geometry as a discipline between the early eighteenth century and the mid-to-late ninteenth century. The most important aspect of this phenomenon for us here is the influence exerted on the practice and conceptualization of geometry as a mathematical discipline. Second, Freemasons created a myth that lay at the foundation of both the educational and professional structures of the newly emerging professional meritocracies of the eighteenth and ninteenth centuries. As such, it is interesting to examine the possible effects this had on the teaching and learning of geometry in these institutions. Third and finally, it is interesting to note the popularity of the movement which used the symbols and mythology, most of which are related to a mathematical discipline. The remnants of this influence are still widely seen in popular culture and the publishing industry, where the concept of 'sacred geometry' in all its forms is a popular topic for those interested in spiritual matters. In this last respect, it should certainly be worthy of note to establish what Masons found was so interesting about geometry and how much they knew of it. This particular aspect of our investigation could give us some understanding of the popularity of some aspects of geometry in contemporary culture.

All of these questions are investigated in this chapter through the analysis of the Masonic link with geometry and the embodiment of Masonic mythology in the life and work of Gaspard Monge (1746–1818), a French mathematician.

Masonic legends and geometry

The greatest aim of the utilization of geometry in Masonic philosophy seems to have been its designation as a model of an ordering principle. Geometrical principles were used to promote activities aimed at establishing an all pervading social order which would, in turn, be a seed of the just society Freemasons set to build. To build, rather than destroy (contrary to what they witnessed around them) was the origin of all philosophy of the Masonic founders.[3]

How this was developed and put into practice can be best seen through a historical analysis. We will therefore first look at the main legends of Freemasonry, their connection with geometry, and the interpretation of the various aspects of geometry through Masonic structure, within its rituals, and in the general symbolism of Freemasonry. All of these various 'applications' of geometry within Freemasonry gave the latter directly, or indirectly, the model for its structure, origin, and purpose.

The structure of Freemasonry – the stone-cutting lodges

The Great Fire of 1666 changed not only the physical but also the social cityscape of London. Until 1666,[4] masons' guilds had an elaborate system of trade control over the whole of the region for which they were responsible. After the Great Fire the need for rebuilding the city far exceeded the skilled labour that was available, and the whole structure of the existing trade controls, especially in terms of entrance to the trade, had to be relaxed in order to attract and maintain a sufficient number of skilled workers. At the same time, the division between two new classes in the building trade itself, that of architect designers and builder-executioners of the designs, became noticeable. When the first ideas of coherent, standardized, and easily replicated designs for houses were brought from Italy to England in the 1670s through Inigo Jones,[5] who studied and developed his own style of Palladian villa, the role of architect as a designer/supervisor was forever divided from the actual execution of the building. As there were no formal or informal qualifications needed to enable one to undertake such profession, it was a question of getting employment in the first place that appeared to matter most. In this regard, social standing and aptitude were essential to an aspiring architect. This was a period of growth for the building trade.

It is unclear how the two processes of building – that of architectural edifices and that of society – became linked in a social organization as they did in Freemasonry. Nevertheless, the first Freemasons seem to have seen the potential in employing the model of the network structure of the stone-cutting lodges to organize their own membership. The stone-cutting lodges had the purpose of regulating and organizing the trade, and their primary business was that of making buildings. The Masonic ideals of society employed the metaphor of the act of building to promote its own social and spiritual aims, and through the system of lodges they were able to develop and spread their ideas. The secrecy that stone-cutting masons used to protect trade secrets, Freemasons used to protect identity of membership; but this seems reasonable having in mind that, after all, the first identified lodge of 'speculative' rather than 'operative' or stone-cutting masons was founded in 1648, during the Civil War, and that members of the lodge included men from both sides of the conflict.

At this point we need to make a distinction between the two systems: that of real, or *operative* stone-cutting masonry, and that of philosophical/social/political, or *speculative* Freemasonry. The first known initiation of a speculative Freemason is well documented: we know much about it as we have records of the initiation of Elias Ashmole (1616–92),[6] which took place in Worcester. A lodge such as that to which Elias Ashmole was initiated came to be known as *speculative* as opposed to the *operative* stone-cutting lodges, its primary aim being philosophical and spiritual speculation rather than the act of building physical structures.

A considerable amount of work has been done by Masonic historians on the growth of Freemasonry in this first period of its official history and the supposed transition from operative into speculative lodges. We will be satisfied to establish only basic facts in this respect. The original lodges were both operative and speculative: when the Grand Lodge of England (the first Grand Lodge in the world) was founded in 1717, the four lodges which constituted it were:

- The Goose and Gridron Ale-house in St Paul's Church Yard lodge,
- The Crown Alehouse in Parker's Lane near Drury Lane lodge,
- The Apple Tree Tavern in Charles street, Covent Garden lodge,
- The Rummer and Grapes Tavern in Channel Row, Westminster lodge.[7]

The first three differed from the fourth in that their memberships were respectively 22, 15, 21, while that of the fourth lodge was 71. It was that fourth lodge, however, that was speculative with only two operative, that is stone-cutting, masons in its ranks.[8] Indeed this lodge counted among its members the Earl of Abercorn, the Duke of Richmond, the Duke of Queensberry, and John Theophilus Desaguliers (1683–1744) who was to become the third Grand Master of the Grand Lodge of England in 1719–20, and was a personal friend (as much as that was possible) of Newton.[9] One can immediately see by the meeting place and the names of these founding lodges that the social aspect was very important if not at the core of the existence of early Freemasonry.

It seems that the structure of the lodges suited this new organization very well. In its first six years, the number of members grew from 129 to 689 and the number of lodges grew from 4 to 56. In these first six years more work was also done on working out the philosophy which would lie at the core of Freemasonry for centuries to come. The task of formulating this philosophy and the purpose of Freemasonry was given to Rev. James Anderson (1680–1739),[10] who was commissioned to provide the history of the Fraternity, while during said process the original records of the four founding lodges disappeared. Instead, what remains are records of Anderson submitting his manuscript at various stages of its completion to the United Grand Lodge of England, and the final version of what become known as the *Constitutions of the Free and Accepted Masons* which was then published in 1723 (Figure 9.1).

This final document consisted of three sections, the first being based on the *Old Masonic Charges*, or documents that originate from the fifteenth century onwards until the founding of the 'speculative' craft.[11] The *Old Charges* were the anchors to the ancient history of the stone-masons' craft and its successor, the Freemasonry. Although Anderson borrowed heavily from these original documents, his version of the *Old Charges* incorporated in the *Constitutions* was not just a series of guild ordinances. Whilst the original documents show no attempt to speak of matters outside the practice of the building trade, Anderson's *Constitutions* make explicit the need to define, preserve, and elevate the moral character of the Masonic membership. Here we find specified the conduct Freemasons should follow, from the behaviour in the lodge to the way they should act with their families or in their immediate social surrounding.

Separate publications relating to initiatory practices were further developed to promote the loyalty of its membership through the creation of steps of advancement, the repeated swearing of allegiance, and the secrecy attached to

Figure 9.1 *The Constitutions of the Antient and Honourable Fraternity of Free and Accepted Masons*, written by James Anderson, under guidance from the Grand Lodge of England. Both the first edition (1723) and the second edition (1738) had a frontispiece by John Pine. The people represented are almost certainly famous Freemasons from the period, and at the basis of their union can be seen the diagram depicting Pythagoras' theorem. Produced with kind permission from the Library and Museum of the United Grand Lodge of England.

the membership and initiatory practices. The latter unified the aim and the form of Masonic lodges and endowed it with a special status.

By the adoption of the lodge structure and the introduction of initiatory practices, members came to believe in their special relationship with the Grand Architect of the Universe. The mythology thus grew of the importance of the ancestry from the operative building craft to the speculative Freemasonry. At the root of this mythology were Euclid and the legendary architect of Solomon's Temple, Hiram Abif.

Euclid and his Masonic rivals

Q: Why was you [sic] made a Fellow-Craft?
A: For the sake of the letter G.
Q: What does that G denote?
A: Geometry or the Fifth Science

(from Pritchard: Masonry Dissected, *London, 1730, p. 12)*

All of the *Old Masonic Charges* upon which the *Constitutions of Freemasonry* were modelled mention Euclid as a founder of a science of geometry, and through it a founder of the Masonic craft. Euclid's treatise on mathematics, *The Elements*, has thus been a permanent feature of Masonic history. The 'worthy Clerk' Euclid shows the way to live a gentlemanly life by the power of the knowledge of geometry; this certainly had a universal appeal and is quite strikingly described in the *Old Charges*:

And the lords of the country (Egypt) grew together and took counsel how they might help their children who had no competent livelihood in order to provide for themselves and their children, for they had so many. And at the council amongst them was this worthy Clerk Euclid and when he saw that all of them could devise no remedy in the matter be said to them "Lay your orders upon your sons and I will teach them a science by which they may live as gentlemen, under the condition that they shall be sworn to me to uphold the regulations that I shall lay upon them." And both they and the king of the country and all the lords agreed thereto with one consent.

It is but reasonable that every man should agree to that which tended to profit himself; and so they took their sons to Euclid to be ruled by him and he taught them the Craft of Masonry and gave it the name of Geometry on account of the parcelling out of the ground which he had taught the people at the time of making the walls and ditches, as aforesaid, to keep out the water. And Isodoris says in Ethomologies that Euclid called the craft Geometry.

And there this worthy clerk Euclid gave it a name and taught it to the lords' sons of that land whom he had as pupils.

And he gave them a charge. That they should call each other Fellow, and no otherwise, they being all of one craft and of the same gentle birth, lords' sons. And also that the most skilful should be governor of the work and should be called master; and other charges besides, which are written in the Book of Charges.[12]

The real identity of Euclid and the *Elements* has been a matter of enquiry for historians of mathematics, the situation best summed up by Itard,[13] who gave three possible hypotheses that historians of mathematics tend to adopt in this regard:

1. That Euclid was a historical character, known as Euclid of Alexandria, born about 325 BC and died about 265 BC in Alexandria, Egypt, who wrote the *Elements* and the other works attributed to him.
2. That Euclid was the leader of a team of mathematicians working at Alexandria around 300 BC, who all contributed to writing *The Elements*, even continuing to write after Euclid's death.

3 That Euclid was not a historical character and that *The Elements* were written by a team of mathematicians at Alexandria who took the name Euclid from the historical character of Euclid of Megara who had lived about 400 BC.[14]

My preference, if belief makes any difference in the historical sense, is that Euclid was a historical character: after all, whilst it seems to us that what he had done was impossibly voluminous, we have examples from the more recent past, for example Euler (1707–83) whose output was on the par with that of *Elements*. (I am pleased to report the anecdotal evidence that most of Euclid's modern compatriots, my Greek colleagues, share this view.)

In Masonic lore the identity of Euclid is not examined: in a sense his existence is so embedded in the *Old Charges* that Euclid becomes one of the three Masonic deities. As such, the name of Euclid can evoke two meanings:

a) the spirit of fellowship and joint enterprise, relating to the Itard's second hypothesis;
b) the spirit of an enlightened being, whose attunement to the knowledge of geometry allowed the betterment of mankind.

The other two Masonic deities are Hiram Abif, the architect of the Solomon's Temple, which models the perfection of a social edifice, and the initiate, the *Universal Freemason*.

Hiram Abif

Hiram Abif, referred to by the Freemasons as the ancient operative Grand Master, worked under Hiram, the King of Tyre. Whilst Hiram the King of Tyre is a biblical figure, Hiram Abif is not mentioned outside of Masonic lore. In the first 13 years of Freemasonry, from 1717 to 1730, there seems to be no mention of this figure until Pritchard (1730) published his *Masonry Dissected* in London. Pritchard's book aimed to expose Freemasons and their rituals as not adhering to 'the ancient tradition', whatever that may have been – and it is unclear what Pritchard thought it was. In fact, there was a political aim behind this publication, which aimed at allowing men of non-gentile background to enter Freemasonry, and with this publication Pritchard created a schism within Freemasonry. Pritchard's intervention ended with the creation of a second Grand Lodge, which gave itself the name of *The Most Ancient and Honourable Society of Free and Accepted Masons*, allowing artisans (rather than only gentlemen and aristocracy) to be more easily admitted. The original Grand Lodge, in response, proclaimed itself as *The Grand Lodge of Moderns* (1751). As the schism was gradually healed (and completely so in 1813), elements of Pritchard's book were adopted by the Masonic establishment itself. Hiram Abif therefore enters Masonic rituals after Pritchard's publication on a regular basis as the Master Mason in the third degree of initiation.

The story of Hiram Abif goes like this. Hiram was a master mason and as such knew secret geometrical knowledge. Apart from the practical geometrical knowledge of stone-cutting masons, there is a reference to a sacred and secret word, the interpretation of which is diverse in Masonic scholarship. Hiram was at one point asked to divulge this secret word, but as he refuses, he is killed and buried by his killers at a secret place. When Solomon[15] hears of the murder of his architect, he orders the execution of his killers and a search for Hiram's body ensues. This story is re-enacted in masonic ritual for the Master Mason, which is a third degree in the Masonic system of initiation, the first two being that of Apprentice and Fellow Craft. In Masonic rituals, various members of the lodge play various roles in it, and so in the third degree the Master Mason is put in the role of Hiram. The ritual goes as follows: when Hiram is found, the Senior Warden of the lodge (a role given to one of the lodge members) applies the grip of the first degree (all three degrees have different types of handshake or Masons' grips), but as the body is highly decomposed it cannot be raised. He then applies the grip of the Fellow Craft, the second degree, but a similar problem occurs. At this point the leader of the lodge, known as Worshipful Master, informs the lodge members that the substitute word for that sacred word, lost by the death of Hiram Abif, will be uttered after his body is raised. The Worshipful Master, who until this time is usually playing the role of King Solomon, reaches down and grasps the hand of the candidate with the grip of the Master Mason, or 'lion's paw' as it is sometimes called in Masonic circles. By this grip the candidate is imbued with living force and raised from the 'dead level to a living perpendicular'.

In the midst of this decomposing imagery of death, geometry is invoked as a life force and one by which the dead can be raised to the 'living perpendicular'. In this story of resurrection, it seems that the two most important elements that make the dramaturgy of the ritual are linked to architecture and geometry. The first is the initiation into the understanding of the role of Master Mason in the preservation of the sacred and secret knowledge. The second element is the geometrical knowledge – by the right fellowship (the right grip), acceptance of one's place in the hierarchy of the lodge (the right degree) and the right conduct – one can become again a 'living perpendicular' which is of course preferable to remaining a rotting cadaver.[16]

Masonic symbolism and geometry

Returning to an all together less disturbing set of images, let us more closely look at a few examples of Masonic symbols that are purely geometrical. The primary few symbols of Freemasonry are certainly the Great Architect of the Universe (also known as G.A.O.T.U.), the square and compasses, and the diagram of Pythagoras' Theorem.

The Great Architect is the Supreme Being in Masonic mythology. Belief in the Supreme Being is not a universally accepted prerequisite for membership in all Masonic systems. England, for example, follows the suit that 'no stupid atheist, or libertine' should be admitted, but the Grand Orient of France,[17] since its 1877 resolution, accepts both libertines and atheists. The square and compasses were adopted as symbols of Freemasonry and refer to the operative stone-masons craft and their knowledge of geometry. Often the capital letter G appears enclosed within the symbol. The 47th proposition of Euclid's *Elements* (Pythagoras' Theorem) is sometimes included in the square and compasses symbol, but also appears in various other ways in Masonic regalia and the physical setup of the lodge for the purposes of ritualistic meetings, for example. The square and compasses, as well as Pythagoras' Theorem diagram, appear as a jewel for a Master Mason (Figure 9.2).

Perhaps more interesting and less static than the individual symbols is the Geometrical Lecture, which was designed for the Holy Royal Arch of Jerusalem degree in English Freemasonry in the eighteenth century. This ritual gives us a fuller insight into the Masonic understanding of geometry as a science as perceived and interpreted by Freemasonry.

This 'side' degree is an extension of the main three degrees mentioned above; prerequisite for admittance to this order is that the candidate is an active Master Mason (so 'graduated' in the third degree) of a Craft Lodge.[18] The date and the origin of this Lecture are not known, but copies remain in Masonic archives and testify to the beliefs and philosophy of a Platonic system in which the five Platonic bodies represent the principles on which the Universe was built. Three questions that are presented to the initiates are related to the nature and origin of the Platonic bodies and their representations within the ritual. The discussion is divided into three parts: mathematical, philosophical, and Masonic (Figure 9.3).

The mathematical discussion, which is presented in some six pages of the lecture, published as a ritual book (4×7.5 cm), begins with an explanation of the simplest geometrical notions: what is plane geometry, what is solid geometry,

Figure 9.2 The Sisson Jewel. One of the earliest examples of a Past Masters' jewel featuring Euclid's 47th Proposition. It was made by the instrument-maker Jonathan Sisson (1690–1747), who was mentioned in Anderson's 1738 *Book of Constitutions* and was a member of one of the original lodges that united to form the Grand Lodge of England in 1717, and that met in the Thatched Tavern, London. Produced with kind permission from the Library and Museum of the United Grand Lodge of England.

etc. The definition of regular solids is followed by a description of mathematical properties of each Platonic solid. This part of the ritual is followed by some philosophical discussion. Platonic thought, which distinguished between sublunary and superlunary worlds – the former imperfect but built upon the five regular 'Platonic' polyhedra, and the latter, perfect, the world of ideas – is obviously taken as the basis for the lecture. Whether the originator of the lecture knew much of Platonic philosophy is not possible for us to discuss here as we have no access to other details about the ritual apart from the ritual itself. But the story contained in it gives us a peculiar mixture of symbolical story and practical guidance on how to identify the process of creation of the Universe and possibly replicate it on this sublunary plane. To become an active builder of the society, one needs to learn the principles upon which the Universe is built. This lesson can then be taken by each individual Mason who would attempt to act

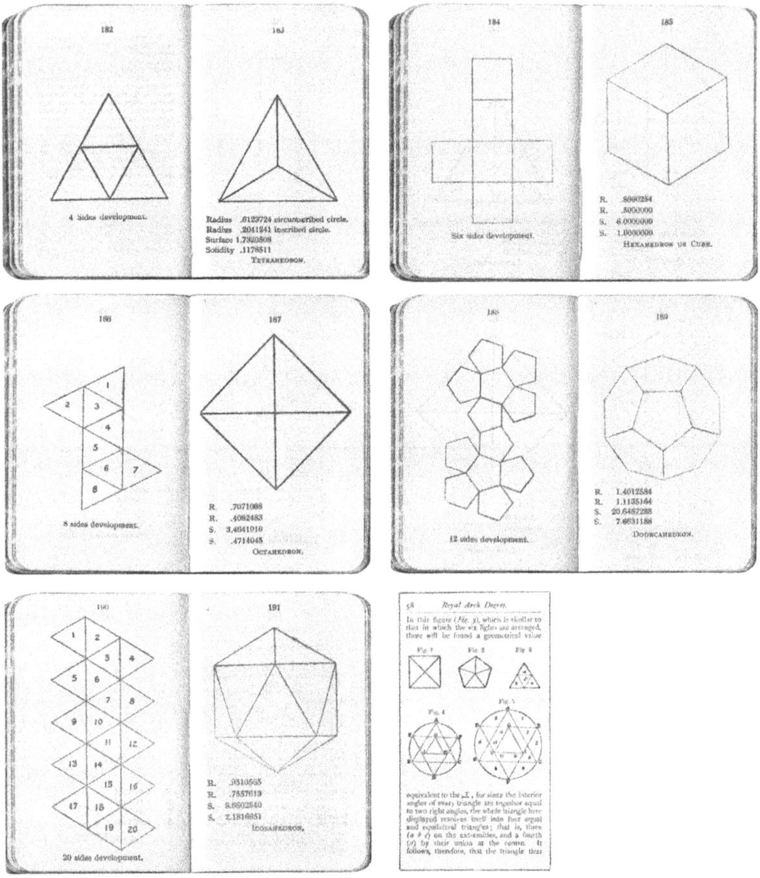

Figure 9.3 Pages from the Royal Arch ritual book from the 1927 edition, *The Perfect Ceremonies of the Supreme Order of the Holy Royal Arch*, published by A. Lewis, London. Produced with kind permission from the Library and Museum of the United Grand Lodge of England.

in his individual sphere of action, and this joint action would eventually create a near-perfect social order through a joint action and through Freemasonry.

Primarily, the aim of this ritual seems to have been to detach a Mason from the considerations of everyday life and raise him to an abstract domain. In that domain the Mason would contemplate first the way the universe was built. The ritual teaches that the four elements – earth, fire, air, and water – were formed out of geometrical elements represented by the forms of the cube, tetrahedron, octahedron, and icosahedron. The dodecahedron is perceived to have been employed by the Great Geometrician in the delineation of the Universe – it is a

Divine Spark, the principle of attraction, and is the ultimate, fifth symbol, the quintessence, by the use of which creation was complete. Much emphasis is put on the fifth element, the attraction principle or the divine spark, noting that this is one of the principles upon which any successful social structure must also rest, as without it the four elements will not spring into life.

Let us now look at how this influenced some mathematicians and how they fitted into the mythology of Freemasonry.

The pioneer of Masonic geometry

Looking at the development of thought in an area of intellectual history, it is tempting to generalize, as in any other area of history. However, it is not easy to identify individuals who lived a life that fits a historical 'movement' so fully as in the case of Gaspard Monge, whose life mirrors so accurately the myth of Freemasonry. As a matter of introduction and contrast to the extraordinary story that belongs to Monge and Freemasonry, it is perhaps prudent to mention a few minor Freemasons and geometers who also embodied Masonic ideals through their personal histories. They come from the British Isles, whilst Monge came from France.

The first would perhaps be too ambitiously called a mathematician. Batty Langley (1696–1751) went from writing books on gardening to inventing his own 'gothic style', and then became a writer of books on practical geometry for artisans and builders in the then growing, and to use the modern phrase, increasingly de-regulated, building trade. At the beginning of his career he wrote mainly on architectural topics, with always the same aim in mind, to familiarize and educate the building craftsmen in the science of geometry. The most important aspect of Langley's work was that it propagated historical interest in the science of geometry amongst the building craftsmen.

Langley discovered the officialdom of Freemasonry in or around 1738 and joined its ranks in the same year. At the same time he seems to have adopted the growing mythology of the brotherhood: that of the line of succession, hidden knowledge, and Freemasons as worthy recipients of such knowledge. It is possible to detect Langley's fervour for a certain Masonic ideal of hidden knowledge, which was alleged to have originated with the ancient Egyptian civilization and to have been passed down through the subsequent centuries to worthy recipients. This thread of 'hidden knowledge' was something he wrote about from the very beginning of his career, but at the time he became a member of the fraternity, Langley attributed this to Freemasonry. The argument of continuity looked rather simplistic, but had (and perhaps still does) a significant influence on the idea of the historical development of geometry and subsequently architecture:

> . . . the Egyptians had handed their secrets to the Greeks; the Greeks devised the orders and passed the secret on to the Romans. The barbarians had wrecked the heritage, but it was rediscovered in the 15th century and given its definite form by Palladio. Inigo Jones was Palladio's true heir . . . There were sceptics, but most people involved in building would have said that however broken, a direct line could be traced from the pyramids to the Banqueting House . . .[19]

The seriousness of Langley's fervour can be seen from his private history (and that of his family) as he called his four sons Euclid, Vitruvius, Archimedes, and Hiram, all according to Langley eminent men from the history of geometry so closely related to the lore of Freemasonry in Langley's writing.

Our second lesser example comes in the form of a mathematician-architect from Scotland. Peter Nicholson (1765–1844) came from a stone-cutting tradition: his father was a bona fide stone-cutting mason from Prestonkirk, East Lothian. Nicholson became interested in geometry and its application to architecture as a young man, possibly because of his background. Nicholson's father George, his uncle Donald, and his brother Hepburn were all members of an operative stone-masons' lodge in Haddington. Peter Nicholson was more ambitious: he went to London to seek his fortune by employing the craft. The Grand Lodge entry describes him as Drawing Master with an address in Newman Street. At this time, Nicholson had run up considerable debts from publishing his book *The New Carpenters' Guide* (1792) as a result of which he went to prison until a new publisher was found who agreed to pay his debt from the proceeds of his new book *Carpenter and Joiners' Assistant* (1797). Nicholson published some further books on practical building and architectural matters; he organized schools in Berwick Street, Soho, during his first period in London, which served as a make-shift institute for mechanics and workmen. At about this time, as he was getting his life back on track in 1798, Nicholson joined Freemasonry. Through it, he made connections with the famous architect brothers Robert and Sydney Smirke (later the architect of the new British Museum) whom he met in the Old Cumberland Lodge in London. There were some other lodges where famous architects and engineers seem to have converged: for example the Jerusalem Lodge no. 197, which met (and still does) at St John's Gate in Clarkenwell, counted among its members the Smirkes, the

Barrys (Charles Barry, 1795–1860, the architect of the new Houses of Parliament; his son Charles Barry 1823–1900; and son of Charles, Charles Edward Barry 1855–1937, all architects and surveyors of Dulwich College Estate), the first professors of architecture at the University College London and King's College London, Thomas Hayter-Lewis and Banister Fletcher (1866–1953) respectively.[20]

Nicholson produced three minor mathematical papers: one on calculus, one on combinatorial analysis, and one on 'involution and evolution'. But his major work was his system of geometrical projection which was used in graphical communication in the British Isles and which was based on his knowledge of the stone-mason's craft and his analysis of Monge's geometrical technique.

Whilst Langley meandered on the outskirts of both geometry and Freemasonry, Nicholson used Freemasonry to employ his knowledge of stonemasonry, and to gain employment with most prominent architectural firms of his time. Monge engaged with both to such an extent that it is sometimes difficult to distinguish Masonic myth from his biography.

Gaspard Monge

Monge was born in Beaune, in May of 1746, the son of a small tradesman. While he worked at the l'École Royale du Génie de Mézières as a draftsman, Monge developed a technique for which he is famous, and which he later called 'descriptive geometry'. Monge became an ardent revolutionary and is credited with being the father of the École Polytechnique. In 1789, after the Bastille was stormed, Monge was appointed the major of the Académie Commission on Weights and Measures that formulated the metric system.[21] Monge also held the position of the Minister of the Navy for a short period, and was sent via Italy with Berthollet (Claude Louis, 1748–1822) on Napoleon's orders as part of the Egyptian expedition. During this expedition, Monge (Figure 9.4) was put in charge of the Engineering corps and although they had met before, became a personal friend of Bonaparte.

Monge's descriptive geometry rested on the tradition of ancient methods of stone-cutting and Monge, as he rose through the ranks of education and taught mathematics around France, began to teach his subject through practical stone-cutting lessons. It is unknown whether Monge would have learnt of stone-cutting techniques through his work, or via the mythology of Freemasonry, but it is certain that he didn't originally conceptualize descriptive geometry in relation to stone-cutting.

Figure 9.4 Gaspard Monge, by Augustin Challamel and Desire Lacroix, in *Album du centenaire, Grands homes et grands faits de la Révolution fraçaise* (1789–1804), published 1889.

Monge conceived his technique while he worked at the engineering school in Mézières when he was given a task to determine the height of the wall in a design of a fortification. Until this time, two methods were mostly used, both involving numerous measurements from the surrounding terrain: one included determining numerous view points from the ground, the triangles determined by the view point, a point of the edge of the fortification, and the height of the wall would all be considered in this process. The other method involved calculations with the heights of each crucial point being measured directly on the terrain and noted on a plan. Both were lengthy and laborious, so Monge's method, which involved much less calculation and could be finished within a few days rather than weeks, was met with suspicion from his superiors.

Instead of measuring many points on the terrain surrounding the plateau where the fortification would have been built, Monge imagined a tangential plane that would determine the height of the wall. This plane is determined by a point and a line: the point lies in the ground plane of the plan of fortification, and the line is the contour of the terrain. Any missile fired at the fortification can be imagined thus to be fired from a surface smoothed by this tangential plane, therefore avoiding the need to measure individual points on the actual terrain (as all points of the actual terrain are either on this tangential plane or below it).[22]

Later on in his career, Monge took this further and studied developable surfaces, the general case of which are tangential planes, which are at the root of his conception of his descriptive geometry.[23]

At first his technique was proclaimed a military secret, as it was developed in a military college, but it nonetheless gave Monge a platform from which his career shot in a pretty much straight and upward manner. At some point though, he connected his invention with the practice of stone-cutting and thus began teaching it in practical settings. His students reported that:

> Instead of making dull details of the stereotomy, our lectures . . . practice the use of the principles involved in the cutting of stones.[24]

Later on, during the Revolution, the École Polytechnique was set up to offer a general education for future public servants and military engineers, and Monge was one of its founding fathers. His technique was established as one of the main subjects taught at the Polytechnique.[25] He saw its role not only in advancing the 'industry' and capability of the nation, but also as a technique through which one could train the mind. Here the geometry was taught in an all together different setting:

> A scholastic discipline which was born in a school, by a school and for a school (but maybe one should say in the École Polytechnique, by the École Polytechnique, and for the École Polytechnique), descriptive geometry allows the passage from one process of training by apprenticeship in little groups which was characteristic of the schools of the Ancien Regime, to an education in amphitheatres, with lectures, and practical exercises, which are no longer addressed to 20 students, but to 400 students. Descriptive geometry also stems from revolutionary methods. A means to teach space in an accelerated way in relation to the former way of teaching stereotomy, an abstract language, minimal, rapid in the order of stenography, descriptive geometry permits a response to the urgent situation as for the education of an elite, which was the case of France at the moment of the creation of the École Polytechnique.[26]

Parallel to this, Monge engaged with metaphorical stone-cutting. From the archives of the Grand Orient of France in Paris, we find that he was the *orateur* (speaker) of the military lodge of the *Corps Royal du Genie* called *l'Union Parfaite* and that he held the degree of the Chevalier d'Orient. Monge also joined the *Ordre des Sophisiens*, and an Egyptian side (not a main degree system) order in French Freemasonry. The development of the Egyptian rite in French Freemasonry was made possible by Count Alessandro di Cagliostro (1743–95), who claimed that this rite held the key to secrets of the philosopher's stone. The philosopher's stone is a mythical stone used in alchemy to transmute matter into gold, but is also related to Freemasonic mythology via Elias Ashmole, the Temple of Solomon, and Platonic philosophy. The premise of its multifaceted supposed meanings relates to knowledge being passed via ancient writings, so that in some legends these writings go as far back as being written by God (or his son) himself. In some alchemical interpretations, the philosopher's stone is ground to be ingested in the search for eternal youth.[27]

Let us at this point look at how the real and the imaginary seem to meet as they do in geometry and in Monge's life: the ideal world of ideas and the applications of geometry to architecture and the world around us. Monge's contribution to geometry stemmed from his work related to stone-masonry (fortification design) and his teaching of the subject was often conducted through practical stone-cutting exercises. His geometrical work was also used and applied in architecture.[28] On the other hand, his involvement with Freemasonry was clearly following the ideals of the social role Freemasons chiseled out for themselves at this time, and his membership of the Fraternity was heavily intertwined with his political activities.

When Monge was admitted to the *Egyptian rite* side degree in French Freemasonry in 1799, Napoleon was about to send a scientific expedition to Egypt to recover the antiquities of interest to the French nation. Monge was put in charge of the Engineering corps. A contemporary of both Monge and Napoleon, and a witness to the expedition, Bourrienne, describes his view of this relationship:

> . . . it was easy to see that he [Napoleon] preferred Monge [to Bertholet, also a member of the Egyptian expedition] whose [Monge's] imagination may have been devoid of precise religious principles, but had a propensity towards religious ideas which was in harmony with Napoleon's own view of this subject.[29]

By a coincidence, and with a further reference to the philosopher's stone, the French expedition came across the Rosetta Stone during this expedition, now one of the British Museum's most prized possessions, which was then considered to be the embodiment of the secret wisdom of the ancient Egyptian civilization. The story of the Rosetta Stone is not the subject of this discourse, but suffice it to say that, whilst the text contained in it nothing that was sacred – it was an official decree issued during Ptolemy V to be placed in public space for all to see – its discovery led to the decipherment of hieroglyphs. This, in turn, eventually led to a deeper understanding of Egyptian culture and civilization, until then largely unknown and mysterious.

Monge's life history follows a rise from a humble background, through the knowledge of geometry, and through the membership of the Masonic fellowship, to the attainment of sacred and secret knowledge. This of course all somewhat collapsed with the fall of Napoleon and Monge's fall from grace towards the end of his life.

Conclusion

One of the most often contemplated questions in modern Freemasonry is that of the meaning of the letter G appearing as a Masonic symbol. *Does the capital 'G' stand for God or geometry?* With this question geometry is positioned against God and elevated to a position that is reserved for deities: Freemasonry de facto elevated geometry to the status of sacred science with this comparison. The various aspects of geometry were thus promoted: geometric diagrams were imbued with 'meaning' through the mythology of symbolism related to them. Through rituals and the artifacts surrounding them, which contained geometric elements, this 'meaning' was internalized by the members of the Fraternity. Furthermore, through the contemplation of how a Mason should conduct his life, geometry was adopted as a metaphorical system which could be analysed, studied, and adopted as a model of moral behaviour to be accepted at all times. 'If in doubt, consult Euclid's *Elements*' can almost be said to be a motto of Freemasonry.

Although the union between geometry and Freemasonry became insignificant after the middle of the nineteenth century, and certainly lost its appeal and credibility in its entirety in the twentieth century, its strength lay in the fact that it became part of the cultural heritage of Western culture. This created the concept of 'sacred geometry' for better or for worse, worse probably being the view of the majority of mathematicians in modern times.

Masonic symbolism related to geometry was a synthesis of historical, mythological, and intuitive sources, producing a set of values that were described in Masonic documents and enacted through rituals and were considered as being universally true and innate. At the very basis of Masonic geometry was the concept of space with its plastic quality – it was a material which the Grand Architect used in the building of the universe. The multiple homologies between the concepts of building the cosmos, society, and one's own personal life and place in society focused on human and social activities related to the craft of building, space as a divine entity, and knowledge of geometry. This placed the Masonic initiate in a context in which space and the craft of building had gained mystical status, but one which an initiate could influence and even control. The imaginary spatial position of the individual was thus promoted from a mere existential plane to the multi-dimensional sphere, from whose centre one was able to focus one's will in order to influence his environment. This was all to be achieved through the knowledge and manipulation of symbols related to the craft of building.[30]

Many of the later nineteenth and early twentieth-century mystical traditions, some stemming from Freemasonry – which can be unpityingly called the Western Mystery Tradition[31] – adopted similar strategies: the principles of individuation through initiation, definition of one's possible sphere of influence, and the identification and later perfection of the technology to do it: in the case of Freemasonry that was the practical craft of building, architecture as a activity, and finally geometry as knowledge of principles. In this way a mathematical discipline found itself in the midst of cultural developments removed from academic or practical uses. Here we revisit one of the questions posed at the beginning of this chapter: the question of the successful means of propagation of the concept of 'sacred geometry' in modern culture. This concept of 'sacred geometry' had a rapid rise through Freemasonry, and persists to our day, independently, as it were, of Freemasonry itself. The quality of the publications thriving on this concept is diverse, with the topics mainly related to geometric constructions and the golden section or divine proportion. Perhaps I should say, however, that the real stuff from which sacred geometry as perceived by Freemasons of the time comes, is not that of the fanciful proportions but the geometry related to stone-masonry. Freemasons admired (and I believe still do) the works of stone-masons built across Europe mainly during the Middle Ages (the golden period of stone-masonry), and glorified both the practical and social skills of stone-masons. Both of these skills are finely synthesized in Monge's work also.

It would now be useful to revisit the remaining two questions listed at the beginning of this chapter.

The first question we asked was to do with the influence of Masonic thought on various aspects of intellectual history related to geometry as a discipline between the early eighteenth and the mid to late ninteenth century. The rise of interest in geometry through Freemasonry and its almost complete identification with the single application – that to the building crafts and professions and their symbolical meaning – had a single most important outcome. This was that geometry became a popular, easily identifiable mathematical discipline with a practical aim and a well-known and identifiable tradition, albeit one relying more heavily on myth than history. The symbolism of Freemasonry made geometry easier to grasp and decipher for a non-mathematical public. After all, it is easier to remember that a set square and compasses are symbols for the sacred act of building and relate this to the Great Architect of the Universe, than it is to understand the actual Pythagorean Theorem and its applications. As a consequence a considerable population from this era learnt of geometry and its sacred symbolism through Freemasonry.

The second issue, that of the role the mythology created by Freemasonry played in the founding of the new schools of learning, is perhaps harder to explain. There is certainly a strong case of circumstantial evidence to show that prominent politicians, statesmen throughout most of Western Europe and the United States were keen and senior members of the Fraternity, although it is also clear that this influence on society did not persist much beyond the mid or late nineteenth century.[32] Freemasonry promoted the powerful belief in the necessity of knowledge of geometry, and further embedded and intertwined this with a pursuit of scholarship. This particular aspect manifested itself in various ways, from the creation of the new institutions of learning – as in the case of Monge in France and the École Polytechnique, and the creation of the University College London founded by leading Freemasons (in full Masonic regalia)[33] at the beginning of the nineteenth century in England – to the founding of 'research lodges' and research libraries within Masonic institutions towards the end of the nineteenth century.[34] However, whilst the fact remains that some important protagonists played a role in both advancing the subject of geometry and establishing the new schools where geometry had a prominent, and very practical role, it would be a far-fetched conclusion to say that as a rule geometry became more important than other subjects in these new institutions of learning.

As a final remark, I want to reiterate that the answer to the question that Freemasonry since its early days has posed – whether capital G stands for geometry or God – is actually irrelevant. The question itself is important: it elevated the status of the science of geometry to that of a Godly science, and the student of that science to that of the 'priestly' being. Monge thus seems like a figure of a perfect priest in Masonic geometry, as in his life and work we find all the elements pertaining to the Masonic myth. He rose from a humble background through the study of geometry, just as the poor sons whom Euclid led to prosperity through the knowledge of geometry in Masonic myth. Most of Monge's work, and certainly his work on descriptive geometry and stone-cutting, can somehow be linked to the ancient knowledge of stone-masons, and his 'fraternal' achievements both professionally and within Freemasonry testify to the strength of the principles the Masonic fraternity held at that time.

Notes and references

1. See C.H. Josten, *Elias Ashmole*, Oxford University Press, 1966; J. Hamill, *The Craft: A History of English Freemasonry*, Crucible, 1986; S. Lawrence, Geometry of Architecture and Freemasonry in 19th Century England, 2002, PhD Thesis, Open University, Milton Keynes, UK.
2. Modern Masonic historians, like Hamill, are keen to promote this vision of Freemasonry. See also Lawrence, Geometry of Architecture.
3. See M. Jacob, *The Radical Enlightenment: Pantheists, Freemasons, and Republicans*, Allen & Unwin, 1991; A. Halpern, *The Democratisation of France, 1840–1901: Sociabilité, Freemasonry and Radicalism*, Minerva Press, 1999.
4. H. Colvin, *1600–1840, Dictionary of British Architects*, John Murray, 1978.
5. Inigo Jones (1573–1652), English architect who built some of the most famous buildings in London.
6. See *Elias Ashmole*, 5 vols, ed. with a biographical introduction by C.H. Josten, Oxford University Press, 1966. Masonic libraries have extensive resources for the study of the origins of Freemasonry, among these there are large numbers of papers related to Elias Ashmole and his initiation. However, the list of papers relating to Ashmole is too large to include in the bibliography. See also *Dictionary of National Biography*, Oxford University Press (1949–50), 1, 644–5.
7. On the history of Freemasonry see Gould, *The History of Freemasonry, Its Antiquities, Symbols, Constitutions, Customs, Etc.*, 6 vols, T.C. Jack, 1884–7; Hamill, *The Craft*, and M. Jacob, *The Origins of Freemasonry: Facts and Fictions*, University of Pennsylvania Press, 2007.
8. 1725 records, the earliest records for this lodge that can be seen in the United Grand Lodge of England's Archive.
9. Hurst's little book on the biography of Desaguliers can be found in some Masonic libraries, see W.R. Hurst, *An Outline of the Career of John Theophilus Desaguliers . . . To which is appended a history of the Edgware Lodge from 1918 to 1928*, printed for private circulation, 1928, London.
10. See J. Anderson, *The Constitutions of the Free-Masons. Containing the History, Charges, Regulations, &c. of that most Ancient and Right Worshipful Fraternity*, Grand Lodge of England, London, 1723; and J. Anderson, *The New Book of Constitutions, of the Antient and Honourable Fraternity of Free and Accepted Masons, containing Their History, etc.* Grand Lodge of England, London, 1738.

11. Freemasonry is often referred to as the 'Craft', as Hamill's *The Craft* testifies.
12. Quote taken from M.S. Cooke. Similar text appears in other copies and manuscripts of the *Old Charges*. See Lawrence, Geometry of Architecture for full list of *Old Charges*.
13. J. Itard and P. Dedron, translated from the French by J.V. Field, *Mathematics and Mathematicians*, Open University Press, 1978 (original 1959).
14. A parallel can be made in this approach to the true story of Nikolas Bourbaki, as André Weil, *Souvenirs d'apprentissage*. Vita mathematica 6. Birkhäuser: Bale, 1991, has described.
15. Solomon, a Biblical figure, also mentioned in other religions such as Judaism, Islam, and Bahá'í Faith, was the son of David, believed to have reigned around 970–931 BC in Israel.
16. Whilst rituals vary across boundaries of cultures and historical periods, the central story is described as above in all surveyed documents, see Lawrence, Geometry of Architecture.
17. It is an accepted fact that the Grand Lodge of England was the first official Masonic Grand Lodge; Grand Orient de France was founded in France in 1773 after the absorption of the older Grand Lodge of France which in turn was founded in 1733. Grand Orient is considered to be mother lodge of most continental Freemasonry. See D. Knoop, *The Genesis of Freemasonry: Development of Freemasonry in its Operative, Accepted, and Early Speculative Phases*, Manchester University Press, 1947.
18. The history of this order is closely related to the schism between two grand lodges which grew out of the schism within the original grand lodge in 1751 mentioned earlier. This degree became very popular in both the Grand Lodge of Moderns and Ancients, and when the two reunited in 1813 remained part of the initiation.
19. J. Rykwert, *The First Moderns. The Architects of the Eighteenth Century*, MIT Press, 1980, p. 197.
20. For details of Masonic membership of these men, see Lawrence, Geometry of Architecture.
21. For Monge's work, see S. Lawrence, Alternatives to teaching space – teaching of geometry in 19th century England and France, in R. Fox and B. Jolly (eds), *Franco-British Interactions in Science Since the Seventeenth Century*, College Publications, 2010, and R. Taton, *L'Oeuvre scientifique de Monge*, Presses Universitaires de France, 1951.
22. S. Lawrence, Developable surfaces: their history and application, in Kim Williams (ed.), *Nexus 2010: Relationships Between Architecture and Mathematics, III* Springer, 2011.
23. See Lawrence, Developable surfaces.
24. As reported in 1777 by J.B. Meusnier (1754–93), one of Monge's most gifted pupils at the school of Mézières. See P.J. Booker, *A History of Engineering Drawing*, Chatto & Windus, 1963, p. 89.

25. Claris Gaston, *Notre École Polytechnique*, Librairies-Imprimeries Réunies, 1895. Ivor Grattan-Guinness, *Convolutions in French Mathematics 1800–1840. From the Calculus and Mechanics to Mathematical Analysis and Mathematical Physics*, Birkhäuser Verlag, 1990. Jőel Sakarovitch, The teaching of stereotomy in engineering schools in France in the XVIIIth and XIXth centuries: an application of geometry, an "applied geometry", or a construction technique?, in Patricia Randelete-Grave and Edoardo Benvenuto (eds), *Between Mechanics and Architecture*, Birkhäuser Verlag, 1995.
26. Sakarovitch, The teaching of stereotomy, p. 211.
27. For alchemical obsessions and their articulation into scientific (in particular chemical and bio-chemical) methods, see M.P. Lawrence, *The Secrets of Alchemy*, Chicago University Press, 2007.
28. J. Sakarovitch, *Epures D'architecture*, Birkhäuser Verlag AG, 1997.
29. J.A.F. Bourrienne, *Mémoires*, 10 vols, Paris, 1831, 1–2: 243.
30. J.P. Slifko, The Ritual Performance and Transmission of the Printed Text in Geographic Place and Across Networks of Place and Space in Early American Freemasonry and Civil Society, 1734 to 1850. PhD Thesis, 2012, UCLA.
31. See Lawrence, Geometry of Architecture.
32. For details on Masonic involvement with the Declaration of Independence in the US, see S.L. Morse, *Freemasonry in the American Revolution*, Masonic Service Association of the United States, Washington, 1924; S.C. Bullock, *Revolutionary Brotherhood. Freemasonry and the Transformation of the American Social Order, 1730–1840*, published for the Omohundro Institute of Early American History and Culture, University of North Carolina, University of North Carolina Press, Williamsburg, Virginia, 1999; and Jacob, *The Origins of Freemasonry.*
33. See Lawrence, Geometry of Architecture.
34. Lawrence, Geometry of Architecture.

CHAPTER 10

Charles Dodgson's work for God

MARK RICHARDS

The book will be a great novelty, and will help, I fully believe, to make the study of Logic far easier than it now is: and it will, I also believe, be a help to religious thoughts, by giving clearness of conception and of expression, which may enable many people to face, and conquer, many religious difficulties for themselves. So I do really regard it as work for God.

(Charles Dodgson, describing his book Symbolic Logic *in a letter to Mrs S.F. Rix, 7 July 1885.[1])*

Charles Dodgson's worldwide fame rests almost entirely on his authorship of *Alice's Adventures in Wonderland* (1865) and *Through the Looking-Glass* (1871), published under the pseudonym 'Lewis Carroll'. In spite of the outstanding success of these books, he devoted his life to researching and teaching mathematics, predominantly at Christ Church, Oxford. The *Alice* books gave him a substantial income, such that he could easily have relinquished his academic duties and survived on his earnings as a children's author. Yet he chose to continue on the career path which he discovered early on at Christ Church and he remained there until he died in 1898. It is clear that he was a competent mathematician and some of his work has been identified as influential. However, the question still stands: why did he take this path and not stray from it in the face of the lure of a good life as an author?

This chapter attempts to explain how Dodgson's religious upbringing and his deep faith in God underpin his life-decisions, and to show that whilst believing writing for children was an important activity, his mathematics was 'work for God' and therefore a more appropriate use of his time and talents. Although one cannot draw direct links between the work he did in mathematics or logic and his faith, in the sense of these sciences challenging or changing his beliefs, it is of interest that in these subjects he found his way to serve God. Moreover, this chapter suggests that many of the questions about Charles Dodgson that biographers and critics have raised over the years can be answered more effectively by examining his mathematics and his religion and the interplay between these aspects of his life, than by simply regarding him as a children's author.

Dodgson's early life – the promising pupil

Born 27 January 1832, Charles Lutwidge Dodgson was the eldest son of Charles and Frances Dodgson and the third of 11 children. At that time, his father was perpetual curate of All Saints Church in Daresbury, Cheshire. Charles Dodgson Senior was a man of significant talent who had attained a double first at Oxford in Mathematics and Classics (Figure 10.1). According to his grandson, Stuart Dodgson Collingwood, he was 'a man of deep piety and of a somewhat reserved and grave disposition . . . tempered by the most generous charity.'[2] However serious and devoted to Christian duty that may make him sound, Collingwood adds that in 'moments of relaxation his wit and humour were the delight of his clerical friends' and paints a picture of a man who might have made a good career as an author or as a mathematician. However, his modesty and firm Christian faith led him to devote his life to using his talents for less celebrated purposes: that is, caring for the local community – tending his flock one might say – looking after and educating the poor, but also pursuing a moderately ambitious clerical career in which he gained a reputation as an effective preacher of sermons and a man of influence. It might be argued that it was his wanting to marry that lead to his relinquishing his academic post in 1827, just two years after receiving the appointment, but his early sermons and strong views on the nature of man's devotion to God suggest that he would have seen a clerical career as a genuine calling.

Daresbury was only a small parish, and with 11 children the family lived a modest life and struggled at times. Promotion was essential for practical reasons. Whilst at Daresbury Charles Dodgson Senior took on the additional role

Figure 10.1 Charles Lutwidge Dodgson's father, Archdeacon Charles Dodgson.

of examining chaplain at Ripon and subsequently was able to move the household to Croft in North Yorkshire, where he held the office of Rector for the rest of his life. He took a keen interest in the young Charles's upbringing: in his education and spiritual guidance. His ambitions for his eldest son would not have meant simply finding an occupation, but a career which made full use of his talents. The father need not have worried, of course, because, as we know, the young Charles proved himself to be talented, hard-working, and devoted to what he believed was God's work.

There are a few accounts of Dodgson's early life at Daresbury and Croft. These show the family engaging in amateur dramatics, writing family magazines to which the brothers and sisters contributed poems and stories, the playing of games, setting and solving riddles; all set against a good traditional approach to education, which involved a considerable amount of religious study. It is hard to find evidence of the *very* young Dodgson's interest or abilities in mathematics. He comes across as intelligent and hard working rather than precocious. There is a story, first told by Stuart Dodgson Collingwood[3] and often repeated, of the young boy finding a book of logarithms and asking his father to explain their use and not being put off by being told that he is far too young to understand them. This is, perhaps, more a sign of persistent curiosity than precocity, but there is clear evidence of his mathematical talents emerging whilst at Richmond School, which he attended from August 1844. In a letter to Dodgson's father, headmaster, James Tate, referred to his passing an 'excellent examination' in mathematics and his 'exhibiting at times an illustration of that love of precise argument, which seems to him natural'.[4] Coincident with this evidence that Charles possessed notable mathematical talents were indications that he had literary ambitions. He wrote numerous poems and short stories, many of which have survived in the family magazines and which display a sense of humour we might today call 'Carrollian'.[5] That is to say, in those early works we can see the emergence of a style of wit and humour which we associate with the author of the *Alice* books.

Dodgson moved to Rugby School in February 1846, where his love of mathematics and his skills in the subject developed. It is also likely that it was at Rugby that he first started to explore his own Christian beliefs. His father was politically and religiously conservative – a champion of the High Church. His beliefs and the sermons in which he expressed them were logically and forcefully stated and might have appeared unquestionable to the younger Dodgson. However, Morton Cohen, in his biography of Dodgson, suggests that at Rugby under the headmastership of Archibald Tait the teenager might have become aware that there were other opinions and been exposed to more liberal views.[6] Although Dodgson did not enjoy his three years at Rugby, he worked hard, was successful, and he impressed his teachers. One teacher, Robert Mayor, said that he 'had not had a more promising boy at his age since he came to Rugby' and Tait expressed his opinion of Dodgson's abilities and high conduct, saying his mathematical knowledge was great for his age and his examination for the Divinity prize was one of the most creditable he had seen. Naturally, the next step was Oxford.

In his father's footsteps – the Oxford undergraduate

In May 1850 he matriculated to become a member of the University of Oxford. This entailed taking examinations as well as swearing to observe the Statutes of the University and signing his name to the Thirty-Nine Articles of Religion. The statutes by then were, largely, a formality, with so many of them being trivial or archaic, but the need to swear allegiance to the 39 articles, which Dodgson would have no trouble doing at that time, is a further indication of the link between education and faith which ran throughout his early life.

Dodgson moved to Oxford in January 1851 (Figure 10.2). His college was Christ Church, just as his father's had been. Unlike his father, however, he remained a member of the college for the rest of his life. In his second year at Christ Church, Dodgson took moderations – the second of the three sets of examinations required for his degree – achieving a first class in mathematics, and a second in classics. As a consequence of this success he was awarded a studentship. Studentships were posts unique to Christ Church – in a sense they were scholarships that entitled the recipient to accommodation in the college and a small stipend giving them the freedom to study without the immediate need to find an income or other sponsorship. Studentships were for life or up to the point where the student married, with the only requirement of note being that the student should prepare for holy orders in the Church of England. That meant first becoming deacon and later a priest – both of which involved examinations and set expectations regarding standards of behaviour. Dodgson's

Figure 10.2 The earliest known photograph of Charles Lutwidge Dodgson (Lewis Carroll); possibly taken in 1855 when he was at an age of 23 years.

father had been a student at Christ Church and took holy orders, but had given up his studentship on marrying. Early letters between the father and the son suggest that this is also what Dodgson senior had in mind for his eldest son. In a sense, this was the usual course of events and to some extent the original purpose behind the studentships – a stepping stone to taking a clerical post. Over the years, the role had become more like a fellowship at other colleges and in time – during Dodgson's lifetime, in fact – the restriction about remaining unmarried was removed and eventually the requirement to become a priest was also relaxed. However, when Dodgson accepted the role he knew that becoming a priest would be part of his planned future. This has become one of the most discussed aspects of Dodgson's life because although he did become deacon he did not proceed to the priesthood. Some have questioned whether he ever fully intended to become a priest, but if that was his intention at first, at what point did he change his mind about this and what were his reasons?[7]

As part of his preparation for his final examinations in mathematics, during 1854, Dodgson spent two months in Whitby in the company of Bartholomew Price at one of his reading parties. Price was Professor of Natural Philosophy at Oxford and a mathematician of note and the role that he played in Dodgson's development as a mathematician was significant. If there was ever a point at which Dodgson could see that the study of mathematics was going to be the basis of his career, then surely it happened during this reading party. Price clearly inspired his students and made an impression on Dodgson. They remained friends right up to the time of Dodgson's death.

Dodgson was awarded his BA, with a first class in mathematics, in December 1854. Always a modest man, but one who was keen to report progress back home to his family, his early letters make fascinating reading. At one point he wrote 'I am getting quite tired of being congratulated on various subjects: there seems to be no end of it. If I had shot the Dean, I could hardly have had more said about it.'[8]

On earning his degree, in a letter to his sister Mary, he wrote:

Enclosed you will find a list, which I expect you to rejoice over considerably: it will take me more than a day to believe it, I expect – I feel at present very like a child with a new toy, but I daresay I shall be tired of it soon, and wish to be Pope of Rome next.

His letter then includes the results from the examination papers followed by:

All this is very satisfactory. I must also add (this is a very boastful letter) that I ought to get the Senior Scholarship next term. [. . .] One thing more I will add, to crown all, and that is – I find I am the next 1st class Math. student to Faussett (with the exception of Kitchin, who has given up Mathematics) so that I stand next (as Bosanquet is going to leave) for the Lectureship. And now I think that is enough news for one post.[9]

The following years were spent settling into post-graduate college life and developing his teaching skills. The next seven years were comparatively uneventful although we do know that he made ambitious plans for further study, he worked hard at mathematics, continued his literary interests, and established himself as a successful amateur photographer.

Post-graduate life at Christ Church and a famous river-trip

Although we might say that, *academically*, Dodgson's early post-graduate years were among his least remarkable, 1861 and 1862 were certainly notable.

In 1861 Dodgson was ordained deacon, and although this first step towards becoming a priest was something he had agreed to take when he accepted the studentship, he still had his doubts about it. These were not religious doubts, but concerns that it might interfere with his chosen occupation of teaching and studying mathematics. Hitherto, his religious commitments and his mathematical work could co-exist without difficulty, but becoming deacon and later, priest, might mean that he would feel an obligation to scale down his mathematical activities or give them up altogether. There have been a number of theories put forward over the years to explain why Dodgson was so reluctant to join the priesthood. In essence these boil down to a simple question that he had to face. How best could he serve God? Nonetheless, he was able to reconcile these conflicts – certainly as far as becoming deacon. One might say that at this point he was managing to keep all his options open. This is all explained rather eloquently in a letter to his cousin William Wilcox, written some years later in 1885:

When I was about 19, the Studentships at Christ Church were in the gift of the Dean and Chapter – each Canon having a turn: and Dr. Pusey, having a turn, sent for me, and told me he would like to nominate me, but had made a rule to nominate *only* those who were going to take Holy Orders. I told him that was my intention, and he nominated me. That was a sort of "condition," no doubt: but I am quite sure, if I had told him, when the time came to be ordained, that I had changed my mind, he would not have considered it as in any way a breach of contract.

> When I reached the age for taking Deacon's Orders, I found myself established as the Mathematical Lecturer, and with no sort of inclination to give it up and take parochial work: and I had grave doubts whether it would not be my duty *not* to take Orders. I took advice on this point (Bishop Wilberforce was one that I applied to), and came to the conclusion that, so far from educational work (even Mathematics) being unfit occupation for a clergyman, it was distinctly a *good* thing that many of our educators should be men in Holy Orders.
>
> And a further doubt occurred. I could not feel sure that I should ever wish to take *Priest's* Orders. And I asked Dr. Liddon whether he thought I should be justified in taking Deacon's Orders as a sort of experiment, which would enable me to try how the occupations of a clergyman suited me, and *then* decide whether I would take full Orders. He said "most certainly" – and that a Deacon is in a totally different position from a Priest: and much more free to regard himself as *practically* a layman. So I took Deacon's Orders in that spirit. And now, for several reasons, I have given up all idea of taking full Orders, and regard myself (though occasionally doing small clerical acts, such as helping at the Holy Communion) as practically a layman.[10]

With matters such as these to contend with, clearly 1861 was a very significant year.

The following year was important for a very different reason. It was 1862 that saw the first telling of the *Alice* stories. It is a digression in this discussion of Dodgson's commitment to carrying out work for God, but it might also be regarded as a digression in Dodgson's own life. On Friday 4 July 1862, Dodgson wrote in his diary:

> . . . Duckworth and I made an expedition *up* the river to Godstow with the three Liddells; we had tea on the bank there, and did not reach Ch.Ch. again until quarter past eight.[11]

'Duckworth' was Robinson Duckworth, an Oxford colleague and 'the three Liddells' were Alice, Lorina, and Edith, daughters of Henry Liddell, Dean of Christ Church. Dodgson had become friendly with the Dean's family and spent considerable time with them. He also took a number of successful photographs of them. Taking the young girls on a boating trip was not such an unusual activity. For two young clergymen like Dodgson and Duckworth, taking young women on such a trip would not be acceptable, but the children of the Dean were young enough to go un-chaperoned and for the trip to be perfectly innocent. It was not an unusual occurrence for Christ Church men to 'borrow' some children for an afternoon's excursion on the river.

The trip they took on 4 July 1862 is of notable importance. On what must have been a magical journey, Dodgson made up a story about a girl, named Alice, and her adventures under ground, to keep the children entertained. On returning to the college at the end of the day, Alice asked Dodgson to write the story down. His initial diary entry does not suggest that he thought the day was particularly remarkable at the time, but some months later he appended the following note to the entry:

> On which occasion I told them the fairy-tale of "Alice's Adventures under Ground," which I undertook to write out for Alice, and which is now finished (as to the text) though the pictures are not yet nearly done. Feb 10, 1863.[12]

Subsequent notes by Dodgson indicate that the pictures for the tale took yet more time to complete, but he was able to present the finished manuscript to Alice Liddell in November 1864. *Alice's Adventures Under Ground* was a beautiful creation – neatly written and illustrated with charming drawings. The manuscript is now in the British Library. The enthusiastic response to the manuscript led him to consider publication. The original story was adapted into the longer and more sophisticated tale that we know today, and published, entirely at Dodgson's own expense, under the title *Alice's Adventures in Wonderland*, in 1865. The original text was expanded from 12,000 words to 27,000, and *Punch* cartoonist John Tenniel was employed to produce the illustrations (Figure 10.3). Being well-written, charming, and ground-breaking, it was hugely successful at the time and has never been out of print since.

Today, an author achieving that level of success might start to give some serious thought to the direction of their career. One might think that in the wake of the success of *Alice*, Dodgson would have spent considerable time working on more children's books or humorous verse, but this was not the case. Given that only a few years earlier he was struggling to make the decision as to whether he could serve God better as a priest or as a mathematical educator, he was unlikely to abandon both options and become a full-time writer for children. However he had acquired many child-friends over the years and he knew the positive effects on a child of a well-told story or some inventive games-playing or puzzle-setting. In *Alice* he had a new tool at his disposal and in spite of his commitment to work at mathematics, he allowed himself time to make further use of his creation.

In 1871 he brought out the sequel, *Through the Looking-Glass and What Alice Found There*. This was a remarkable achievement as it is rare for a sequel of any book to be as much loved as the original. It is also interesting to observe how alike the two books are in many respects, yet how totally different they are in others. It is clear that he put a lot of work into making *Looking-Glass* as polished

Figure 10.3 'Alice accepting a thimble from Dodo' by John Tenniel from *Alice's Adventures in Wonderland.*

a work as *Wonderland* and made sure it was not simply a case of publishing more of the same.

He also found many other ways to expand the popularity and scope of 'Alice'. For example, in 1889 he brought out a simplified version of *Wonderland* for much younger readers, called *The Nursery 'Alice'* and even published a facsimile of the original manuscript of *Alice's Adventures Under Ground*. There was also a stage version, the adaptation and production of which he oversaw to some extent, and a few pieces of what, today, we would call merchandise, such as card games, a case for storing postage stamps, and a decorated biscuit tin.

In spite of these digressions, which occurred from time to time following the success of *Alice*, it is clear that he never let his most successful creation distract him to any great extent from what he saw as his real work.

Finding the right path

"Would you tell me please, which way I ought to go from here?"
"That depends a good deal on where you want to get to," said the Cat.[13]

There is no doubt that the success of *Alice* had a strong influence on the rest of Dodgson's life. It brought him a good income, a little, perhaps not always welcome, recognition, and it gave him a new means of communication. The fame and popularity of 'Lewis Carroll' was quite an asset and one he was often willing to exploit. From time to time, articles, pamphlets, and letters to newspapers would be signed 'Lewis Carroll' rather than Charles Dodgson, especially if it might get a wider audience. These are not simply instances of a famous children's author bringing out something new to please his audience or earn some income, but examples of a man with particular views using the opportunity to educate or to influence public opinion.

In 1867, Dodgson sent a short children's story called *Bruno's Revenge* to Mrs Gatty, editor of *Aunt Judy's Magazine*. Mrs Gatty responded enthusiastically, lavishing praise on Dodgson's talent as a writer:

> You may have great mathematical abilities, but so have hundreds of others. This talent is peculiarly your own, and as an Englishman you are almost unique in possessing it. If you covet fame, therefore, it will be (I think) gained by this.[14]

We do not know Dodgson's response to this, but it is not unreasonable to imagine that he would have been horrified by it. Least of all by the logical contradiction of suggesting that the talent is 'peculiarly his own' and then saying that he is 'almost unique' in possessing it. There is sufficient evidence in Dodgson's diaries and letters to show that he would not have coveted fame at all and whatever his abilities were at writing fairy stories, this would not have been a worthy career for a man who had other talents. He certainly had a desire to produce amusing pieces like *Bruno's Revenge* for children, but this was clearly a matter of light relief from his work as a mathematician at Oxford and not something he regarded as essential in his own life. In fact we can see that his mathematical work around that time was actually making great steps forward.

Prior to the success of *Alice*, Dodgson's printed works on mathematics are mainly teaching aids, demonstrating how he was getting his curriculum and methods in order. Although we know him to have had mathematical skills that went beyond teaching, there is little in the way of published work from that time which can be regarded as having a wider significance. After the publication of *Alice* there was a flurry of work which made a notable contribution to the subject and is today regarded as having historical importance. Now in his mid-30s he appears to be finding a clearer direction for his work. It is in the years immediately after *Alice* that we start to see Dodgson producing some significant publications.

During the 1860s, geometry had been his principle area of work and this was a period during which Dodgson published at least three important pieces among others. *Condensation of Determinants* (1866) publishes his method of calculating determinants of larger order matrices by a method which steadily reduces the size of the matrix to the simple two-by-two case. The following year, he published his *Elementary Treatise on Determinants* and in 1868 he produced *The Fifth Book of Euclid Treated Algebraically*. Whilst the late 1860s and 1870s saw Dodgson committing his time to geometry, he gradually started to explore other areas of mathematics. In the 1870s he worked on voting procedures, and explored number theory (Figure 10.4). By the 1880s he was making significant steps towards what would become one of his major projects – his textbooks on symbolic logic.

In 1881 he gave up the post of Mathematical Lecturer. He no longer needed the income it brought for him and in a sense the time was more valuable than the money. In his diary, he wrote that his resignation was so he could spend more time 'partly in the cause of Mathematical education, partly in the cause of innocent recreation for children and partly, I hope (though so utterly unworthy

Figure 10.4 Charles Lutwidge Dodgson, c. 1874 at an age of around 42 years.

of being allowed to take up such work) in the cause of religious thought'. Giving up the lectureship was by no means an act of retirement. In the concluding poem from *Through the Looking-Glass* he wrote:

Ever drifting down the stream
Lingering in the golden gleam
Life, what is it but a dream?

Dodgson never allowed his life to drift and rarely lingered on anything. He had lists of projects he wanted to complete and was a man full of ideas and ambitions. Contrary to what has been suggested by some writers, he led a very busy social life.[15] He had his child-friends whom he took to art galleries and theatres and he had many adult friends. He socialized and he took part in all aspects of college life. He was a busy man, but one might say that he engaged in too many different activities. If he really wanted to produce the best possible work, he needed to abandon a few. As his life progressed he did reduce the number of social engagements, eventually gave up photography (which had been a time-consuming hobby) and giving up the lectureship was part of the process of refocusing his attention.

The matter of priesthood had not gone away. He had not completely ruled it out, but Christ Church seemed to have come to terms with the idea that he could retain the studentship even if working towards becoming a priest was not being carried out in earnest. It is possible that he still saw it as a back-up career, although clearly one which would have presented some difficulties. Whilst I have given strong emphasis to the idea that his main objection to a clerical career was that he might see it as distracting him from his mathematics, there were other issues. His strong interest in the theatre, for example, which was frowned upon by the Church, and his own views about how religious services should be performed might have caused him difficulty. Both of these matters we will come to shortly, but we must also consider the fact that he had a stammer. There are differing views about how serious this was, but it is clear that reading texts would have been difficult for him – and, no doubt, he felt that it might undermine the sentiment of a bible-reading during a service if his congregation were distracted by the lack of fluidity in his speech. Preaching was not such a difficult task, because he was able to choose his words carefully to avoid the awkward consonants and combinations of them. However, reading a lesson from a printed text was always prone to difficulties.

He carried on with his mathematics and logic, his story-telling and games and puzzles for children, and he also contributed to the cause of religious thought as he declared he would, but in his own particular way. The latter took the form of sermons – often to groups of children – and written advice to family and friends seeking spiritual guidance. The help that he gave to people on religious matters was always carefully thought out, argued logically, and expressed with eloquence. During this time, he developed his own views on how services should be conducted and what constituted acceptable behaviour in the eyes of God. In a letter from 1885, he suggests that his views were still broadly in line with those he first inherited from his father:

> . . . I myself belong to the "High Church" school. My dear father was a "High Church" man, though *not* a "Ritualist," and I have seen little cause to modify the views I learned from him, though perhaps I regard the holding of different views as a less important matter than he did.[16]

In spite of stating here (and elsewhere) that his views had changed little from his father's, there are indications in his letters and diaries that he became slightly more liberal as he came to terms with his own religious beliefs. His emphasis that his father was not a ritualist is of note. Dodgson strongly disliked any form of ritual in church services and chose the churches he attended quite carefully. In some cases his views were, for the time, rather controversial. He thought it perfectly acceptable, for example, for young children to read or even engage in gentle play during church services, if they were not old enough to fully comprehend them. Significantly, he thought what was acceptable in his own behaviour was a matter for his own conscience. In a letter from May 1892 he wrote:

> The main *principle*, in which I hope all Christians agree, is that we ought to abstain from *evil*, and therefore from all things which are *essentially* evil. This is one thing: it is quite a different thing to abstain from anything, merely because it is *capable* of being put to evil uses. Yet there are classes of Christians (whose *motives* I entirely respect), who advocate, on this ground only, total abstinence from
>
> (1) the use of wine;
> (2) the reading of novels or other works of fiction;
> (3) the attendance at theatres;
> (4) the attendance at social entertainments;
> (5) the mixing with human society in any form.
>
> All these things are *capable* of evil use, and are frequently so used, and, even at their best, contain, as do *all* human things, *some* evil. Yet I cannot feel it to be my duty, on that account, to abstain from any one of them.[17]

Dodgson's careful use of the words 'some' and 'all' here are of note. By 1892, when the letter was written, he was heavily engrossed in his work on symbolic logic, much of which concerned the syllogism. A syllogism is a pair of statements and a conclusion which can logically be derived from them. Such statements were generally expressed in forms like 'All humans things contain evil' and part of the reason why Dodgson worked on symbolic logic was to teach people how to avoid the traps of jumping from a particular assertion ('some . . .') to a universal conclusion ('all . . .'). Dodgson stops short of turning his argument into a series of syllogisms, but careful reasoning of this kind was clearly something on which both his work on logic and his religious discourse were founded.

However, in spite of having these clearly thought out and elegantly expressed views, as he grew older and worried about how much time he had left at his disposal, he did not feel compelled to work towards publishing his projected book on religious difficulties. Instead, his work on symbolic logic expanded and his conviction of its importance grew.

'Work for God'

Dodgson's early printed works on logic were examination papers and examples of problems to be worked out in class. Believing that he needed to find new ways of teaching elementary reasoning, he invented his 'Game of Logic,' which used a board and counters to solve syllogisms. The success of this as a teaching aid led him in 1886 to publish his first book on the subject – *The Game of Logic*. The book was published under the name Lewis Carroll, because he believed it would gain better attention and he was committed to encouraging more people to take up the subject.

So convinced was he of the benefits of learning logic that he allowed the subject gradually to take up most of his time. The main exception was that whilst in the midst of all this logic work he published his only other sizable children's stories – *Sylvie and Bruno* and *Sylvie and Bruno Concluded*, but he never really spent as much time on them as they required. The books contain some of his most original writing, collected together over a number of years, but rather than spend the necessary time carefully crafting a great work, he simply threaded together this collection of disparate pieces using a rather naïve love story. It is hard to believe that this could ever have been a successful venture, but its failure, if that is not too strong a word, can be attributed to the fact that he was unable to find the time needed to make it work. The beauty of the *Alice* books is that they are written in a way that they can be fully appreciated by children and adults alike. The tragedy of the *Sylvie and Bruno* books is that adults find much of them too childish and children find them, on the whole, too adult. In Dodgson's defence, he had other, more important, work to do and would have found it hard to justify spending the necessary time to make *Sylvie and Bruno* more successful. The reason he allowed himself to be distracted from his work on logic at all, is possibly that he saw this book as an opportunity to preach, in a gentle way, to his child readers.

Today logic is predominantly regarded as part of the foundation of mathematics or as a component in the design of computers and electronic circuits. In this sense the subject changed quite dramatically in the years after Dodgson's death. Dodgson, however, saw the study of logic as an exercise in reasoning (as Euclid's geometry was often regarded) in the belief that it would help young people to think clearly and to draw sound conclusions from the facts with which they were presented. In particular, it would help anyone struggling with religious questions to reason them out soundly. Happy with the response to his small book on his method of solving syllogisms, *The Game of Logic* (Figure 10.5), he continued to work on a projected three volume textbook called *Symbolic Logic* (Figure 10.6). This would start by covering the ground of the smaller earlier work, but expanding it to cover more complex logical problems. The second volume would further develop these ideas to cover the ground of the standard logic textbooks of the day. A third volume would explore a host of advanced logical ideas and curiosities. Again, the works were to be published under the

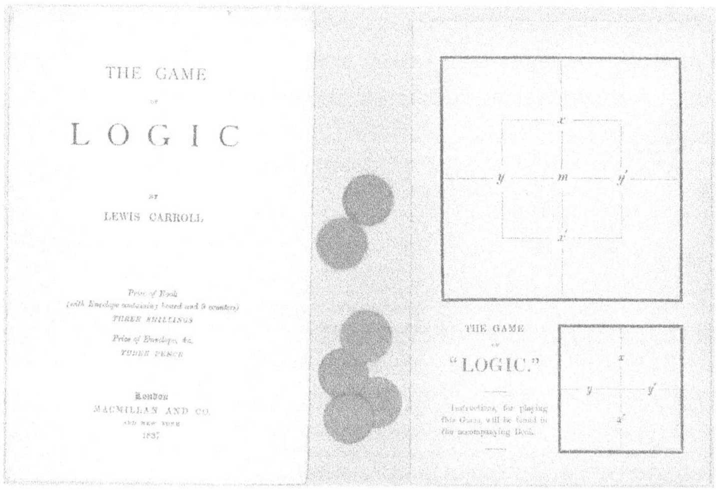

Figure 10.5 Card and counters for use with *The Game of Logic*.

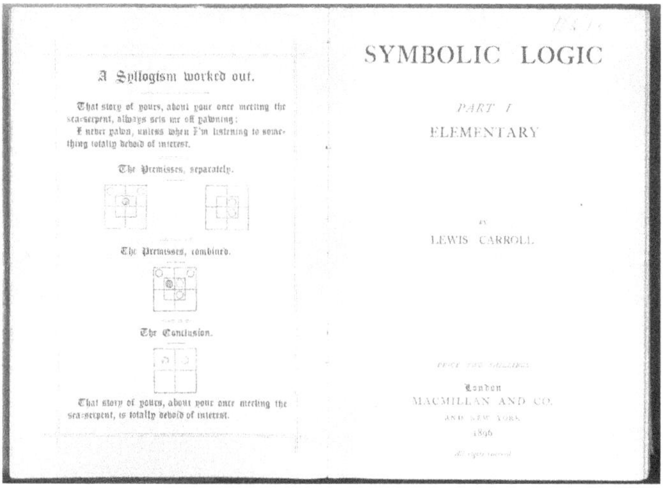

A Syllogism worked out.

That story of yours, about your once meeting the sea-serpent, always sets me off yawning;
I never yawn, unless when I'm listening to something totally devoid of interest.

The Premisses, separately.

The Premisses, combined.

The Conclusion.

That story of yours, about your once meeting the sea-serpent, is totally devoid of interest.

SYMBOLIC LOGIC

PART I

ELEMENTARY

BY

LEWIS CARROLL

London
MACMILLAN AND CO.
AND NEW YORK
1896

Figure 10.6 Frontispiece and title page from *Symbolic Logic*.

name 'Lewis Carroll' in order to gain a wider readership than if they had been published under his real name.

With quite a publishing history behind him and no requirement, either financially or academically, to produce more mathematical works, one might be curious to know why he devoted an ever increasing amount of time to this particular project, at this stage in his life. In a letter to his sister Louisa, written in September 1896, whilst in Eastbourne near the end of the long vacation, he explains just how important he saw this particular piece of work.

I am beginning to realise that, if the books I am still hoping to write, are to be done at all, they must be done now, and that I am meant thus to utilise the splendid health I have had, unbroken, for the last year and a half, and the working-powers, that are fully as great as, if not greater than, what I have ever had. I brought with me here the MSS, such as it is (very fragmentary and unarranged) for the book about religious difficulties, and I meant, when I came here, to devote myself to that; but I have changed my plan. It seems to me that that subject is one that hundreds of living men could do if they would only try, much better than I could, whereas there is no living man who could (or at any rate who would take the trouble to) arrange, and finish, and publish, the 2nd Part of the Logic. Also I have the Logic book in my head: it will only need 3 or 4 months to write out; and I have not got the other book in my head, and it might take years to think out. So I have decided to get Part II finished first: and I am working at it, day and night. I have taken to early rising, and sometimes sit down to my work before 7, and have 1½ hours at it before breakfast. The book will be a great novelty, and will help, I fully believe, to make the study of Logic far easier than it now is: and it will, I also believe, be a help to religious thoughts, by giving clearness of conception and of expression, which may enable many people to face, and conquer, many religious difficulties for themselves. So I do really regard it as work for God.[18]

Symbolic Logic Part 1 came out in 1896 and ran to four numbered editions – 2000 copies in total. When Dodgson died in 1898 parts two and three of *Symbolic Logic* were still unfinished and his 'fascinating mental recreation for the young', as he described it in an advertisement for the book, never really took off.

Dodgson left behind a mass of material, notes, and printers' proofs, from which it is possible to get a rough idea of the direction he was working in, but not enough of the advanced material survives for us to get a true picture of how his work *might* have been perceived had it been published.[19]

The majority of the mathematical subjects that Dodgson tackled have either virtually disappeared from the curriculum – Euclid's geometry for example – or they have changed dramatically in their use or purpose, such as symbolic logic. As a result it is problematic to assess Dodgson's contribution to the history of mathematics, but scholars continue to make progress in this subject and his reputation as a mathematician is growing. Dodgson's success as a mathematical educator, though, is clear and beyond question. Through his games, puzzles, and highly original texts he offered fresh approaches to teaching mathematics and demonstrated that the rigours of the subject make its study an important constituent of everyone's education in life.[20] For all the pleasure and influence that his *Alice* books have generated, his study of mathematics and his teaching of it became his true calling. He saw it as his duty to work in that discipline and he saw the effects of his work as contributing to the greater good.

Whether Charles Dodgson senior ever felt his son had made the *best* use of his talents, we cannot tell. He died in 1868, three years after the publication of *Alice* and more or less around the time that Dodgson was beginning to make serious strides in his work. So it would be nice to think he saw that his son was on the right path. In a sermon dating from 1837, Dodgson's father stated that:

The deference universally paid to splendid talents and the eminence to which they often so rapidly attain, is a proof of their comparative rarity, and a proof, therefore, that the great mass of good must be produced by the aggregate force of those whose single efforts may seem to be of little value and efficacy.[21]

Dodgson must, surely, have done his duty on both counts. Today, we can regard him as being a splendid talent – the word genius is sometimes used to describe him as a writer – yet his devotion to his work for God did not mean that his genius prevented him from being part of that aggregate force to which his father referred. It could be argued that if the father had been able to see the son's life's work he would have approved. His old tutor, mentor, and friend, Bartholomew Price, on the other hand, *was* able to appreciate Dodgson's life's work. Price outlived Dodgson by a few months and on reading his obituary in *The Times* sent a short note to Dodgson's siblings.

> . . . I feel his removal from among us as the loss of an old and dear friend and pupil, to whom I have been most warmly attached ever since he was with me at Whitby, reading mathematics, in, I think, 1853 – 44 years ago! And 44 years of uninterrupted friendship . . . I was pleased to read yesterday in *The Times* newspaper the kindly obituary notice: perfectly just and true; appreciative, as it should be, as to the unusual combination of deep mathematical ability and taste with the genius that led to the writing of 'Alice's Adventures'.[22]

There have been numerous biographies of Charles Dodgson and their authors frequently attempt to answer questions such as why he did not proceed to Priest's Orders, why he did not marry or why he did not write a third 'Alice' story. Generally, these biographers set about writing the life-story of the man who wrote the *Alice* books – and not the biography of a Victorian clergyman or an Oxford mathematician. Although there is, now, a significant body of work that assesses Dodgson's contributions to mathematics, very few writers have attempted to explain the importance of mathematics in the man's life. Robin Wilson, who also contributes to this volume in Chapter 5 on Renaissance combinatorics, is almost alone in treating Dodgson as, primarily, a mathematician, in his *Lewis Carroll in Numberland*.[23] Even the more significant and analytical biographies appear to shy away from the mathematics, although some, such as Cohen,[24] are willing to explore Dodgson's religion in depth. However, that still remains an area waiting further analysis. The approach of regarding religion and mathematics as less important in Dodgson's life than his authorship of the *Alice* books often leads to a distorted view of the man as child-obsessed, dull in everyday life, and socially inept.

If we begin our attempts to understand Dodgson by exploring his religious convictions and his commitment to teach and study mathematics, a very different man emerges and more reasonable answers to those important questions can be found. It is in the interplay between mathematics and religion in Dodgson's life that we can find the key to unlocking his character and explanations for the actions of his life. Moreover, truly to understand the man we must refrain from seeing him as a children's author and, instead, observe him as a man committed to carrying out 'work for God'.

Notes and references

1. Reprinted in: *The Letters of Lewis Carroll*, edited by Morton N. Cohen, Macmillan, 1979, Vol. 1, pp. 585–6.
2. Stuart Dodgson Collingwood, *The Life and Letters of Lewis Carroll*, T. Fisher Unwin, 1898, p. 8.
3. Collingwood, *The Life and Letters of Lewis Carroll*, pp. 12–13.
4. Collingwood, *The Life and Letters of Lewis Carroll*, p. 25.
5. Some of these early works have been published, notably: Lewis Carroll (ed.), *The Rectory Magazine*, University of Texas Press, 1975; Lewis Carroll, *The Rectory Umbrella and Mischmasch*, Cassell & Company, 1932. In the preface to *Mischmasch*, the last of these to be produced, Dodgson describes seven 'domestic magazines' of which he claims to be author or editor of four.
6. Morton Cohen, *Lewis Carroll, A Biography*, Macmillan, 1995, p. 347.
7. For example, Dodgson's 'qualms' about entering the priesthood are discussed extensively in Cohen, *Lewis Carroll*.
8. Letter to his sister Elizabeth, 9 December 1852. Reprinted in Cohen, *The Letters of Lewis Carroll*, p. 22.
9. Letter to his sister Mary, 13 December 1854. Reprinted in Cohen, *The Letters of Lewis Carroll*, Vol. 1, pp. 29–30.
10. Reprinted in Cohen, *The Letters of Lewis Carroll*, Vol. 2, pp. 602–3.
11. The private journals of Charles Dodgson, published as: *Lewis Carroll's Diaries*, edited by Edward Wakeling in 10 volumes. The Lewis Carroll Society, 1993–2007, Vol. 4, pp. 94–5.
12. Wakeling, *Diaries*, Vol. 4, p. 95.
13. From *Alice's Adventures in Wonderland*, Chapter VI.
14. Reprinted in Collingwood, *The Life and Letters of Lewis Carroll*, p. 109.
15. Karoline Leach (*In the Shadow of a Dreamchild*, Second Edition, Peter Owen, 2009) carefully dispels the myth that Dodgson was reclusive and even goes as far as to say that 'he was almost addicted to company'.
16. Letter to Mrs S.F. Rix, 7 July 1885. Reprinted in Cohen, *The Letters of Lewis Carroll*, Vol. 1, pp. 585–6.
17. Letter to A.R.H. Wright, 12 May 1892. Reprinted in Cohen, *The Letters of Lewis Carroll*, Vol. 2, p. 902.
18. Letter to Mrs S.F. Rix, 7 July 1885. Reprinted in Cohen, *The Letters of Lewis Carroll*, Vol. 1, pp. 585–6.
19. During the 1960s and 1970s William Warren Bartley III searched for and collated printer's proofs, manuscripts, and letters written by Dodgson which would reveal what might have been published in the second and third volumes of *Symbolic Logic*.

Bartley's reconstruction of those volumes, with comprehensive notes was published as *Lewis Carroll's Symbolic Logic*, Harvester Press, 1977.

20. The originality that Dodgson applied to teaching symbolic logic is matched in his writings on several other branches of mathematics. *A Tangled Tale* (Macmillan, 1885) is a collection of mathematical puzzles embedded in a series of linked short stories. *Euclid and his Modern Rivals* (Macmillan, 1879) is a humorous, but serious, discursion, in the form of a stage play, on the nineteenth-century authors of textbooks on Euclidean geometry. See Robin Wilson's *Lewis Carroll in Numberland* (Penguin Books, 2008) for a comprehensive study of Dodgson's work in mathematics, games, and puzzles.
21. Rev. Charles Dodgson, *A Sermon Preached in the Minster at Ripon on Sunday, January 15, 1837*. Published by the Diocese of the Lord Bishop, 1837. I am grateful to Charlie Lovett for bringing this rarely seen sermon to my attention.
22. Reprinted in Collingwood, *The Life and Letters of Lewis Carroll*, p. 353.
23. Wilson, *Lewis Carroll in Numberland.*
24. Cohen, *Lewis Carroll.*

CHAPTER 11

P.G. Tait, Balfour Stewart, and *The Unseen Universe*

ELIZABETH F. LEWIS

Introduction

In April 1875 an anonymous publication appeared, in which the possibility of immortality and the existence of an unseen universe were argued for on a scientific basis. The authors – revealed to be two major figures in nineteenth-century physical science – believed that they had found unity in the latest scientific theories and the established doctrines of Christianity.

The authors approached their task humbly, as 'reverent students of the Scriptures', hoping to communicate that science, properly understood, does not stand in opposition to religion.[1] Plainly, in their own words: 'Our object, in this present work, is to endeavour to show that the presumed incompatibility of Science and Religion does not exist.'[2] Guided by their fundamental Principle of Continuity, they believed that they could show science to be, in fact, the 'most efficient supporter' of Christian doctrines.[3]

They aimed their argument at those for whom scientific objections to man's immortality and the existence of an invisible world – raised by some in the scientific community – were proving to be insurmountable hurdles to a belief in these doctrines.

The anonymity game

The Unseen Universe; Or, Physical Speculations on a Future State ran through three anonymous editions before authorship was revealed officially in April 1876. With each edition came the critics' reviews and conjectures on the mysterious authorship issue.

Publishing anonymously using the third person plural prompted many to assume a joint or collaborative effort. Rumours of two distinguished physicists were supported by evidence from the book of the authors' 'thorough knowledge of the highest and most recent developments of natural philosophy'.[4] To some, however, the book's title and the nature of some of the authors' hypotheses were suggestive of spiritualist authors.

In May 1875 *The Spiritualist Newspaper* was able to quash rumours of spiritualist authors, by reporting the *Athenaeum*'s revelation of the authors' identities: 'Dr Balfour Stewart, of Manchester, and Mr P.G. Tait, Professor of Natural Philosophy at the University of Edinburgh.'[5] In the preface to their fifth edition, Tait and Stewart admonished the 'London "Weekly,"' who stated their names as facts, and did so without authorization, barely days after publication of the first edition.[6] Other critics played fairer and refused to publish names, in respect of the authors' desire to remain anonymous.

Despite such an early outing, which no doubt came as a blow to Tait and Stewart, there were pockets of individuals who remained unaware of the authors' true identities and the curious among them continued to search the text for clues left unwittingly by the authors. Naturally, those familiar with Tait and Stewart's scientific interests and their connections within the scientific community would have easily deduced authorship. There are references to Sir William Rowan Hamilton (1805–65)[7] and William Thomson (1824–1907),[8] later Lord Kelvin, who were both associated with Tait. There are detailed references to sunspots and J.D. Forbes' work on glaciers, which both relate to Stewart. There are also many explicit references to Tait and Stewart's earlier publications, for instance: Tait's *Thermodynamics*, Tait and Thomson's *Natural Philosophy*, Tait and Steele's *Dynamics of a Particle*, and Stewart's *Conservation of Energy*. References to the experiments carried out jointly by Tait and Stewart, on a disk rotating *in vacuo*, are also very suggestive.

As late on as December 1875, William Thomson's name remained associated with the book: as one of three authors (Tait, Thomson, and Stewart) or as half of the already known collaborative team, Tait and Thomson. Tait responded to this succession of candidates by annotating a review published in *The Glasgow Herald*, with a reference in Greek to Book VI of Homer's *Iliad*: 'φυλλωνγενεη'.[9]

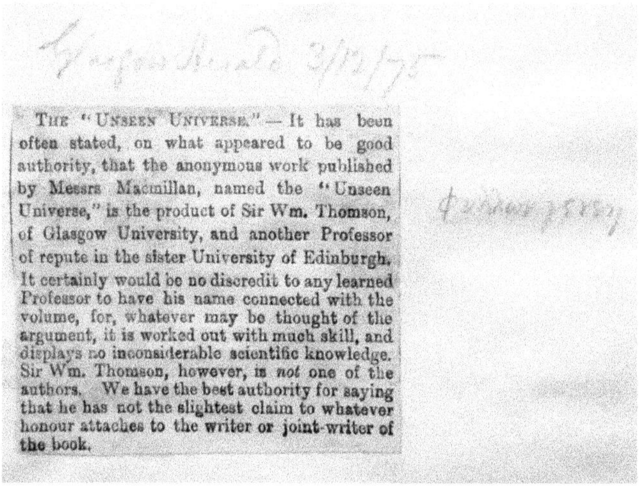

Glasgow Herald 3/12/75

THE "UNSEEN UNIVERSE."—It has been often stated, on what appeared to be good authority, that the anonymous work published by Messrs Macmillan, named the "Unseen Universe," is the product of Sir Wm. Thomson, of Glasgow University, and another Professor of repute in the sister University of Edinburgh. It certainly would be no discredit to any learned Professor to have his name connected with the volume, for, whatever may be thought of the argument, it is worked out with much skill, and displays no inconsiderable scientific knowledge. Sir Wm. Thomson, however, is *not* one of the authors. We have the best authority for saying that he has not the slightest claim to whatever honour attaches to the writer or joint-writer of the book.

Figure 11.1 Tait's annotation of a review published in *The Glasgow Herald*, 3 December 1875. Sourced from Tait's scrapbook in Edinburgh. Reproduced with the kind permission of the James Clerk Maxwell Foundation.

See Figure 11.1. I quote from an 1866 translation by the English mathematician and astronomer, Sir John F.W. Herschel (1792–1871):

Man's generations flourish and fall, like the leaves *of the forest.*

Leaves on the earth by winds are strown, yet others succeed them,

Ever renewed with returning spring. So fares it with mortals:

One generation decays and its place is filled by another.[10]

The consensus view as reported by *The Glasgow Herald* was that Thomson and 'another Professor of repute' had co-authored *The Unseen Universe*, which was contradicted by the newspaper, who claimed to have it on 'the best authority' that Thomson was not one of the authors.[11]

The original newspaper cutting is preserved, along with over 40 other reviews and documents relating to *The Unseen Universe*, in Tait's scrapbook – a collection of miscellaneous papers, newspaper cuttings, letters, and so on, which once belonged to Tait. The scrapbook is preserved by the James Clerk Maxwell Foundation, at the birthplace of James Clerk Maxwell, 14 India Street, Edinburgh.

It is interesting to speculate on Tait and Stewart's reasoning behind choosing anonymity. By withholding authorship, their readers would approach the work without preconceived ideas of what they might expect from a well-known, named author: the book would be judged solely on the quality of the authors' speculations. Anonymity would also serve as a tactical measure, to promote discussion which would draw in a readership. Since Tait and Stewart invited criticism of their hypotheses and since they dealt with the criticism they received robustly and publicly, it is inconceivable that they opted for anonymity in order to shield themselves from harsh criticism and to protect their reputations. Tait, certainly, was not adverse to the cut and thrust of public debate.

Evidence exists which suggests that Tait, in particular, was prepared to resist the draw of recognition so that he might enjoy participating in the intellectual speculations surrounding anonymity. In May 1875 a critic from *The Nation* newspaper brought to light a communication that had been published in *Nature* magazine on 15 October 1874. The communication, signed 'West', gave a proposition in the form of an anagram:

$A^8\ C^3\ D\ E^{12}\ F^4\ G\ H^6\ I^6\ L^3\ M^3\ N^5\ O^6\ P\ R^4\ S^5\ T^{14}\ U^6\ V^2\ W\ X\ Y^2$

With its letters correctly arranged it reads: 'Thought conceived to affect the matter of another universe simultaneously with this may explain a future state.' The critic recognized *The Unseen Universe* as 'the full elucidation and expansion' of this proposition and went on to reason that West was in fact 'one of the two reputed authors of "The Unseen Universe," and presumably the senior partner'.[12]

Again a cutting of the review is preserved in Tait's scrapbook. Figure 11.2 is a photograph of the original. The name West has been circled and a hand-written

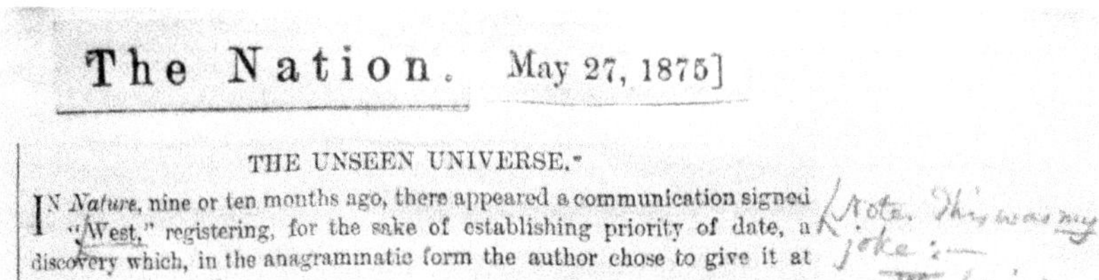
The Nation. May 27, 1875]

THE UNSEEN UNIVERSE.*

IN *Nature*, nine or ten months ago, there appeared a communication signed "West," registering, for the sake of establishing priority of date, a discovery which, in the anagrammatic form the author chose to give it at that moment, ran thus :

$A^8\ C^3\ D\ E^{12}\ F^4\ G\ H^6\ I^6\ L^3\ M^3\ N^5\ O^6\ P\ R^4\ S^5\ T^{14}\ U^6\ V^2\ W\ X\ Y^2$.

These letters placed rightly compose a proposition—"Thought conceived to affect the matter of another universe simultaneously with this may explain a future state"—of which this book is the full elucidation and expansion. It seems probable, therefore, that "West" was one of the two reputed authors of the 'Unseen Universe,' and presumably the senior partner. (Of him it may be said that he is at all events no sciolist in physics, and the way in which Biblical texts and the most ruthless of modern scientific hypotheses combine to shape and support his conclusions is, in the year of grace 1875, a phenomenon which really gives one an impression of freshness and originality.)

Figure 11.2 Tait's annotation of a review published in *The Nation* newspaper, 27 May 1875. Sourced from Tait's scrapbook. Reproduced with the kind permission of the James Clerk Maxwell Foundation.

annotation, signed by Tait, has been inserted: 'Note. This was my joke:- We (viz) S & T!' So the name West stood for both authors, 'Stewart & Tait', who had written the communication in *Nature* so as to cryptically reveal authorship well in advance of publication and thereby establish priority when the first edition appeared.[13]

Peter Guthrie Tait (1831–1901): a brief biographical sketch

Peter Guthrie Tait (Figure 11.3) was, in his time, 'one of the most renowned scientists and mathematicians in Europe'.[14] Few could claim a greater reputation

Figure 11.3 Peter Guthrie Tait: an etching by William Brassey Hole, 1884. From the 'Quasi Cursores': the gallery of portraits of the Principal and Professors of Edinburgh at the time of the Tercentenary in 1884. ©National Portrait Gallery, London. Tait is depicted explaining the fundamentals of electrostatics. He is operating the Holtz machine, which converts mechanical energy into static electricity by a process of induction. The static electricity produced is stored in the Leyden jar capacitors, also shown. The formula $Q = CV$ gives the voltage-charge relationship for the capacitor: Q is the charge on the capacitor; C is a constant called the capacitance and V is the voltage across the capacitor. The formula $W = \frac{Q^2}{2C}$ gives the energy stored in the capacitor, W, in terms of charge and capacitance.

than Professor Tait: for his researches in physical science (experimental and theoretical) and in mathematics; for his contribution to the Royal Society of Edinburgh and for his powers of exposition in the lecture theatre. Beyond the nineteenth century, however, Tait has been unfortunately overlooked – overshadowed, perhaps, by the brilliance of his personal friends, James Clerk Maxwell (1831–79)[15] and William Thomson.

Tait was born in Dalkeith, Midlothian on 28 April 1831, to parents John Tait, Secretary to the Fifth Duke of Buccleuch, and Mary Ronaldson, daughter of John Ronaldson (a tenant farmer). He was educated at the celebrated Edinburgh Academy – where he established a life-long friendship with fellow student, James Clerk Maxwell – and at the Universities of Edinburgh and Cambridge. At Edinburgh he studied mathematics under Philip Kelland (1808–79)[16] and natural philosophy under James David Forbes (1809–68).[17] In 1852 he received his BA from Cambridge, graduating as Senior Wrangler (the second Scot on record) and First Smith's Prizeman (MA 1855).[18] A fellowship at Peterhouse kept him in Cambridge until 1854.

From Cambridge Tait went to Belfast as Professor of Mathematics at The Queen's College, where he enjoyed a number of happy and profitable associations.[19] Through Thomas Andrews (1813–85), Vice-President and Professor of Chemistry at Queen's, he was introduced to experimental work and to William Rowan Hamilton, the inventor of the ingenious method of quaternions.[20] After Hamilton, Tait is considered to have been the leading expounder of the quaternionic theory and the foremost advocate for its use in physics.

Tait returned to Edinburgh in 1860 – with a wife, Margaret Archer Porter (1839–1926) and a daughter, Edith – to take up the chair of natural philosophy at the university there, succeeding Forbes who had gone to St Andrews as Principal of the United College.[21] Tait's biographer, C.G. Knott (1856–1922) described Tait as being 'unsurpassed' as a lecturer.[22] Knott was a former student of Tait's at Edinburgh and his assistant in the natural philosophy department between 1879 and 1883. In 1861 Tait was elected a Fellow of the Royal Society of Edinburgh.[23]

The end of Tait's long and remarkable career followed shortly after the death of his son Freddie in February 1900. Lieutenant Freddie Tait (1870–1900), of the 2nd Battalion of the Black Watch, was killed in action in the South African War.[24] Tait was profoundly affected by the tragic event and from the time of his bereavement his own health began to fail. In December 1900 he stopped teaching at the university and on 30 March 1901 he formally retired from the natural philosophy chair. He died four months later on 4 July 1901.

Amongst Tait's chief contributions to mathematics was his work on quaternions and knot enumeration. In the area of physical science: he devised the first thermoelectric diagram; he invented a pressure-measuring instrument called the Tait Gauge for use in investigations relating to the *Challenger* Expedition;[25] and, encouraged by William Thomson, he turned his attention in later years to the kinetic theory of gases, giving the first proof of the Waterston–Maxwell equipartition theorem. A keen golfer, Tait engaged in pioneering work on golf ball aerodynamics. He is best remembered, however, as the co-author, with William Thomson, of the epoch-making *A Treatise on Natural Philosophy*.[26, 27]

Aspects of Tait's character often led him into bitter disputes with other leading scientists. Knott believed that it was Tait's straightforward approach and his loyalty to his friends and colleagues that often led to his involvement in controversy – 'always on behalf of others'.[28] He entered public disputes with the Irish natural philosopher John Tyndall (1820–93) over priority in energy physics and Forbes' glacier work. Over Thomson's thermodynamic discoveries he came into conflict with the Prussian, Rudolf Clausius (1822–88). And his intense dislike of the vector calculus that threatened Hamilton's quaternions brought him into conflict with the Englishman Oliver Heaviside (1850–1925) and the American J. Willard Gibbs (1839–1903).

On the spiritual aspect of Tait's life, Knott offers the following insights:

> Tait was indeed a close student of the sacred records. The Revised Version of the New Testament always lay conveniently to hand on his study table; and frequently alongside of it lay the Rev. Edward White's book on Conditional Immortality. I am not aware that he distinctly avowed himself a believer in this doctrine, as Stokes did, but he often expressed the high opinion he held of Edward White and his writings. His reverence for the undoubted essentials of the Christian Faith was deep and unmovable; and nothing pained him so much as a flippant use of a quotation from the Gospel writings.[29]

Robert Flint (1838–1910), Professor of Moral Philosophy at St Andrews (1864–76) and Professor of Divinity at Edinburgh (1876–1903), described Tait as someone who had managed to reconcile in his own life the scientific and religious aspects:

> Our departed friend had no sympathy with theological dogmatism, and little with anti-religious scepticism, and consequently held in contempt discussions on the so-called incompatibility of religion and science. At the same time, he had a steady yet thoughtful faith in God, and in that universe which no mere eye of sense, aided by any material instrument, can see. This faith must have made his life richer, stronger, and happier than it would otherwise have been.[30]

Tait was a member of the Scottish Episcopal Church, while Stewart was baptised into the presbyterian Church of Scotland.[31]

Tait's acquaintance with Balfour Stewart began in 1861, when Stewart – a former student of, and lab assistant to, Tait's predecessor Forbes – was appointed an additional examiner in mathematics for the University of Edinburgh. The pair collaborated from 1866 onwards, initially working together on experiments investigating the heating of a disk rotating *in vacuo*.[32]

Balfour Stewart (1828–87): a brief biographical sketch

Balfour Stewart (Figure 11.4), a distinguished Scottish physicist and meteorologist, is remembered for his dedication to observation and experimental research; and for his ability to make from these incisive deductions in the areas of radiant heat and solar phenomena in particular.

Figure 11.4 Balfour Stewart. From E. Youmans and W. Youmans (eds), Dr Balfour Stewart. *The Popular Science Monthly*. New York: Appleton and Co.; 1877; 11: p. facing 257. Original source unknown.

Stewart was born in Edinburgh on 1 November 1828, to parents William Stewart, a tea merchant of Leith, and Jane Clouston, daughter of William Clouston (a minister of Stromness, Orkney). Stewart's university education began, at the age of 13, at St Andrews. From St Andrews he went on to Edinburgh where he studied natural philosophy under Forbes. Encouraged by his parents, he embarked on a career as a merchant; but after a brief period spent in Leith and Australia, he returned to Edinburgh to resume his interest in physical science.

In February 1856 he was appointed as assistant observer to John Welsh (1824–59) at Kew Observatory and in October of that year, assistant to Forbes at Edinburgh.[33] In 1859 he succeeded Welsh as Director of Kew Observatory. In 1862 he was elected a Fellow of the Royal Society of Edinburgh; Honorary Fellowship followed in 1878.[34] Stewart remained at Kew until 1870, when he took up a position as Professor of Physics at Owens College, Manchester.[35] He died in 1887, while on holiday in Drogheda, Ireland. He had married in 1863 and had three children.

Stewart might have had a distinguished career as a mathematician – he had shown promise as a mathematician under the influence of Kelland at Edinburgh and had published a paper on the theory of numbers – but he chose to devote himself to experimental science.[36] His interests – particularly in heat, meteorology, and terrestrial magnetism – were influenced by his association with Forbes.[37]

Stewart's most important contribution was his work on radiant heat: in Forbes' laboratory in Edinburgh he undertook researches which led him to an extension of Prévost's Law of Exchanges. For these researches he was awarded the Rumford Medal by the Royal Society of London in 1868. Tait explains the significance of Stewart's contribution:

> His paper (which was published in the *Transactions of the Royal Society of Edinburgh*) contained the greatest step which had been taken in the subject since the early days of Melloni and Forbes. The fact that radiation is not a mere surface phenomenon, but takes place like absorption throughout the interior of bodies, was seen to be an immediate consequence of the new mode in which Stewart viewed the subject.[38]

Stewart was the author of a number of other noteworthy papers, including those in which he presented the results of his experiments on: radiant light emitted through glass, tourmaline, and uniaxial crystals; and, as noted before, with Tait, the heating of a disk rotating *in vacuo*.[39]

At Kew, Stewart was actively engaged in research into meteorology and terrestrial magnetism. Under his guidance Kew became a centre for the standardization and testing of the instruments used in experiments in these fields.[40] In the area of solar physics he investigated the connections between sun spots, planetary configurations, and terrestrial meteorology.

At Owens College Stewart had as his students the Nobel Prize winner and discoverer of the electron, J.J. Thomson (1856–1940), and the distinguished physicist, John H. Poynting (1852–1914).

Little information is available on the religious aspect of Stewart's life. The best there is in this regard is the following description of him:

A devoted and fervent Churchman, who in later years was a member of a committee appointed by Lambeth Conference to promote interchange of views between scientific men of orthodox opinions in religious matters, he maintained throughout his career a deep interest in the more mysterious problems of existence, and became one of the founders of the Society for Psychical Research, over which he presided from the year 1885 until his death.[41]

The Christian man of science

As Christian men, the authors of *The Unseen Universe* believed that they encountered God both in the study of nature and in the Scriptures. For them there was no conflict between these two sources: they believed them to be of the same divine origin and to constitute the two Books of Revelation. Referring to the Book of Nature metaphor, Tait and Stewart write: 'In fine, the physical properties of matter form the alphabet which is put into our hands by God, the study of which will, if properly conducted, enable us more perfectly to read that Great Book which we call the Universe.'[42]

Drawing on these two sources, Tait and Stewart would inevitably face criticism. Their appeal to Scriptures to unite physical theories and the spiritual was regarded by many as unscientific; they were criticized on points of religious licence and accused of venturing beyond their own areas of expertise. One critic applied the warning of the German botanist Schleiden (1804–81) who was one of the first to accept Darwin:

The first rule which the exact investigator of nature should observe is, that he should not allow himself to pronounce an opinion, either in affirmation or negation, on subjects which do not fall and cannot fall within the sphere of his observation and experience . . . If the natural philosopher comes, not in his special capacity, but in that of man merely, to speak of these matters (as every man has a right to do), then he must have before his eyes the second rule, which is, that he must not pass opinion, form his judgement, nor utter it, upon matters of any science to the present level of which he has not brought himself.[43]

To those who accused Tait and Stewart of 'invading the province of religion' they responded:

we do not write for those who are so assured of the truth of their religion that they are unable to entertain the smallest objection to it. We write for honest inquirers—for honest doubters, it may be, who desire to know what science, when allowed perfect liberty of thought and loyally followed, has to say upon those points which so much concern us all.[44]

Consistent with Christian teachings, our authors had faith in a Creator who created us in His own image so that we might be capable of knowing Him. This doctrine encourages an honest inquiry into those fundamental questions which demand both a scientific and theological approach. With an intimate knowledge of Christian doctrine, acquired through personal faith, and with a profound understanding of contemporary science, few would have been better placed than Tait and Stewart to ponder these questions and to identify the 'connecting link between Revelation and Science'.[45] Their scientific investigations, functioning as an inlet to Revelation, might even be considered to constitute religious service.

Tait and Stewart expected opposition to the theories which they were putting forward, yet they invited reactions from within both the scientific and religious camps: 'Entertaining these views we shall welcome with sincere pleasure any remarks or criticism on these speculations of ours, whether by the leaders of scientific thought or by those of religious inquiry'.[46] A bold invitation indeed.

The science-versus-religion debate: Tait and Stewart's contribution

The Unseen Universe was regarded by some at the time as a model of how to engage in the science-versus-religion debate – which was at its height in that era – 'according to the properly exacting conditions of Science'.[47] In their own minds, the authors' particular contribution to the debate was to identify, once and for all, the true antagonists.

Tait and Stewart's hypotheses were intended to demonstrate that there is nothing inherent in science that cannot accommodate Christian doctrines. Accomplishing this, materialism might be exposed as the true foe of religion. Tait, in particular, expressed a fervent dislike of materialists, referring in one instance to the 'ignorance' which shows itself in the 'pernicious nonsense of the Materialist'.[48] Clearly, he thought materialism abhorrent and a dangerous deviation from real truth.

On this basis, it was inevitable that readers of *The Unseen Universe* would make the connection with John Tyndall's Belfast address, which Tyndall had delivered in August 1874 as President of the British Association for the Advancement of Science.[49] Although Tait and Stewart made no reference to the address and no explicit reference to Tyndall, it was determined that *The Unseen Universe* had been written as a refutation of Tyndall's address.

Tyndall's Belfast address

John Tyndall was a self-professed materialist, who argued, with characteristic directness, for the freedom of scientific inquiry from religious authority and an end to religious intrusion into the domain of science; maintaining the 'superior authority of science over religious or non-rationalist explanations'.[50] His philosophy was shared by fellow members of the X-Club; a group of friends and eminent scientists who formed themselves into a London society. One of its members, the mathematician Thomas A. Hirst (1830–92) explained: 'the bond that united us was devotion to science, pure and free, untrammelled by religious dogmas'.[51]

The theme of Tyndall's Belfast address was the historical development of man's intellect, with an emphasis on how well he realizes the natural impulse, inherent in men, to consider the 'sources of natural phenomena'.[52] Primeval man, rightly and naturally, drew on his experiences but erring in his focus – looking to man and not to nature – his theories took on an 'anthropomorphic form' so that 'supersensual beings . . . were handed over the rule and governance of natural phenomena'.[53] From the relationship between these capricious gods and mankind developed a sub-theme in Tyndall's address – the association of religion with fear, superstition and restricted freedom. Whenever in the course of the history given by Tyndall, scientific development is found to be slow, halting or practically non-existent, the Christian influence is blamed.

Some modern commentators have preferred to see Tyndall as a misunderstood pantheist rather than a materialist.[54] They have associated with him a love of nature and a belief in a Power which is 'immanent in or identical with the universe', according to the definition of pantheism.[55] Whatever the label, it is clear from his address that Tyndall's views would have challenged the traditional doctrines of the Christian faith, for instance his insistence on a materialistic explanation for the universe and the origin of life.

Giving an account of Darwin's theory of evolution, Tyndall considered the implications of Darwin's primordial germ – the common origin of all life. His conclusion: if we are to abandon the idea of creative acts and say instead that the primordial germ developed from matter, then a new and very different understanding of matter is required, since the traditional conception cannot admit of life coming out of matter. It was a real frustration for Tyndall that science was unable to prove experimentally that life can develop from anything other than life. Keen to apportion blame for science's inability to understand the relationship between matter and life, he found fault with the first to define matter – the mathematicians.

While Tyndall insisted on the necessity of materialism, he accepted its insufficiency: 'Understanding' alone cannot satisfy man and for this reason, 'physical science cannot cover all the demands of his nature'.[56] Therefore, 'Understanding' must be supplemented, with: passion, 'Awe, Reverence, Wonder'; 'love of the beautiful, physical, and moral, in Nature, Poetry, and Art' and 'religious sentiment'.[57] Tyndall's use of the term 'religious sentiment', which is loaded with offence, no doubt served to distinguish between faith, and philosophical or scientific reasoning.

Tyndall was prepared to tolerate religious sentiment but he mistrusted its development into doctrine and religious practice; but even these might be permitted, so long as they adapted to fall in line with all other evolving forms of knowledge: 'The facts of religious feeling are as certain to me as the facts of physics. But the world, I hold, will have to distinguish between the feeling and its forms, and to vary the latter in accordance with the intellectual condition of the age.'[58]

Undoubtedly, Tyndall's address had the potential to offend and antagonize Christian scientists who maintained traditional beliefs. The biographers of William Robertson Smith (1846–94), who Tait and Stewart consulted in the course of putting the book together, testify to this. Smith was a Scottish theologian and Semitic scholar and, between 1868 and 1870, assistant to Tait at Edinburgh. His involvement in *The Unseen Universe* is evidenced in a series of letters he received from Tait.[59] His role was that of consultant: he was asked to give his opinion on proofs and to suggest improvements. His biographers recalled:

Tyndall's address created a sensation both in the theological and in the scientific world which was quite out of proportion to its importance as a serious attack on the orthodox position. It gave special offence to a distinguished group of scientific men who, like Lord Kelvin and Clerk Maxwell and their great predecessor, Faraday, were staunch upholders of the truths of revealed religion. This feeling of irritation was probably the immediate occasion of *The Unseen Universe*, a work of some celebrity in its day, which may be regarded as an elaborate counterblast to Dr Tyndall's provocative manifesto.[60]

A further connection between Tyndall's address and *The Unseen Universe* is the correspondence in themes. Both touch upon: the continuity of nature; the practice of reaching beyond the bounds of the senses with intellect; a history of atomic theory; the mind–body problem; molecular processes and consciousness; and the mysterious link between matter and life, and energy and life.

Tait's biographer, Knott, established the chronological link between the two:

In the winter of 1874, a few months after the delivery by Tyndall of his famous presidential address before the British Association at Belfast, it began to be whispered among the students of Edinburgh University that Tait was engaged on a book which was to overthrow materialism by a purely scientific argument. When, in the succeeding spring, *The Unseen Universe* appeared it was at once accepted as the fulfilment of this rumour.[61]

All this suggests that *The Unseen Universe* was written as a refutation of Tyndall's address. But while we might admit that the book was hastily compiled in order to produce a timely response to Tyndall, we should realize that much of the substance of *The Unseen Universe* is present in the authors' earlier work. Tait and Stewart themselves anticipated accusations alleging a hasty, ill-thought-out contribution. In the preface to the first edition: 'the ideas here developed ... are not the result of hasty guessing, but have been pressed on us by the reflections and discussions of several years'.[62]

The Principle of Continuity

The book's most difficult concept to understand is the Principle of Continuity – the thread off which all the arguments hang.[63] The authors promise to define it but never do so explicitly – an illustration in terms of astronomy has to suffice – and each time the Principle is invoked there is some re-modelling of it. So we appeal to the critics for clarification, who in their frustration went to great efforts to formulate a precise definition.

The interpretation which is closest to what the authors appear to have had in mind offers a dual definition. In the spirit of science, the Principle constitutes a belief in the uniformity of the laws of nature: 'The government of the universe has proceeded on a certain plan, ruled by certain fixed laws, we may therefore infer that it will continue to be so.'[64] In the spirit of religion, the Principle is an expression of trust: 'God has endowed us with certain capacities which enable us to dwell safely in the world and serve Him according to His laws. He will not distress or alarm His children by capriciously suspending or setting aside the laws which guide His universe.'[65] Another critic, in effect, unites the two definitions by describing the uniformity of natural law as 'the steady expression of the unchanging Will of the Creator'.[66]

In the preface to the fourth edition, the authors (who felt some clarification was warranted) asserted that the Principle 'has solely reference to the intellectual faculties'.[67] So perhaps the best interpretation of the Principle is as some sort of intellectual process – a process which makes sense of a 'continuous chain of cause and effect, of antecedent and consequent'.[68]

Tait and Stewart were more explicit about what constitutes a breach of continuity: 'Continuity, in fine, does not preclude the occurrence of strange, abrupt, unforeseen events in the history of the universe, but only of such events as must finally and for ever put to confusion the intelligent beings who regard them.'[69]

We will encounter the authors' application of the Principle as we follow their arguments. Tait and Stewart proposed the following working hypotheses on the basis of the latest scientific investigations and discoveries. These hypotheses attempt to explain the beginning and end of the universe, and the production of life, and to realize the possibility of immortality.

The authors' hypotheses: the Great First Cause, the beginning of the universe, and the origin of life

From the outset, Tait and Stewart identify themselves as believers in a Creator God: 'Let us begin by stating at once that we assume, as absolutely self-evident, the existence of a Deity who is the Creator of all things.'[70] Still, they are men of science and as such they believe that they are required to adopt a particular philosophy when speculating on the origins of natural phenomena:

> We think it is not so much the right or privilege as the bounden duty of the man of science to put back the direct interference of the Great First Cause—the unconditioned—as far as he possibly can in time. This is the intellectual or rather theoretical work which he is called upon to do—the post that has been assigned to him in the economy of the universe.
>
> If, then, two possible theories of the production of any phenomenon are presented to the man of science, one of these implying the immediate operation of the unconditioned, and the other the operation of some cause existing in the universe, we conceive that he is called upon by the most profound obligations of his nature to choose the second in preference to the first.[71]

Maintaining this philosophy throughout, Tait and Stewart consider first the beginning of the universe, and the origins and evolution of life.

Following the development hypothesis of Kant and Laplace they explain how the universe was formed: initially there was a 'diffused or chaotic' state; then, when gravitating matter condensed and coalesced, potential energy was converted into heat and visible motion; and at the centre of what would become our solar system, a swirling mass threw out satellites and 'planetary attendants' as it cooled.[72] Thus the process of evolution had two principle elements – the integration of matter and the dissipation of energy.

When the earth had matured sufficiently to produce favourable conditions, the first forms of life appeared and, from these, complex biological life-forms developed. In formulating their hypotheses on the origins of life, Tait and Stewart do not offer any scientific objections to Darwin's primordial germ: they cannot, of course, deny the staggering explanatory power of Darwin's hypotheses. Still, they have to account for its existence.

They assume that this first form of life requires a living antecedent; for life cannot develop from anything other than life and the act of its creation would constitute a breach of continuity. And this antecedent must be conditioned, for the Principle of Continuity requires 'an endless development of the conditioned'.[73] To be 'conditioned' is to be subject to the laws of the universe – 'laws according to which the beings in the universe are conditioned by the Governor thereof, as regards time, place, and sensation'.[74]

Rejecting the hypotheses of abiogenesis and spontaneous generation, Tait and Stewart turn to Scripture for the conditioned, living antecedent of Darwin's primordial germ:[75]

If we now turn once more to the Christian system, we shall find that it recognises such an antecedent as an agent in the universe. He is styled the Lord and Giver of Life. The third Person of the Trinity is regarded in this system as working in the universe, and therefore in some sense as conditioned, and as distributing and developing this principle of life, which we are forced to regard as one of the things of the universe, in the same manner as the second Person of the Trinity is regarded as developing that other phenomenon, the energy of the universe.[76]

In their eagerness to provide cohesive and comprehensive hypotheses, Tait and Stewart are somewhat presumptuous in their interpretations of the Christian God and the specifics of Trinitarian doctrine. Having said this, the routes to this hypothesis are well sign-posted: the Holy Ghost is the Spirit residing in the souls of the faithful, who works in preparation for everlasting life; the Son is the developer of the Will of the Father which is expressed in the laws of the universe in which energy plays a fundamental role.

In accounting for the variety of species, Tait and Stewart rule out separate acts of creation – maintaining the scientific man's philosophy. Informed by Darwin, Huxley, and Wallace, they cite natural variation, and natural and artificial selection as the probable causes. Still, the initial creative force remains the same: 'We have driven the creative operation of the Great First Cause into the durational depths of the universe,—into the eternity of the past,—but for all that we have not got rid of God.'[77]

The authors' hypotheses: the end of the visible universe

Scripture and Science both point to a coming catastrophe, the one in language a child can understand, the other in the wordless eloquence of Nature's changeless laws.[78]

Central to the science behind Tait and Stewart's grand scheme is the objective existence of matter and energy. The objective existence of matter is a conviction based on the conservation of matter; a law which enshrines the 'experimental truth' that matter is not susceptible to changes in quantity.[79] Tait and Stewart reason that if we admit the objective existence of matter, then we must afford an objective existence to anything which is conserved '*in the same sense*' as matter.[80] Examining a number of possibilities from abstract dynamics, they find that energy is alone conserved in this very particular sense: while energy may undergo transformation into a variety of forms, the sum of all the various energies in a closed system remains constant, according to the conservation of energy.

The characteristic natures of matter and energy are described in a novel fashion in the following quotation from the book:

> matter is always the same, though it may be masked in various combinations, energy is constantly changing the form in which it presents itself. The one is like the eternal, unchangeable Fate or *Necessitas* of the ancients; the other is Proteus himself in the variety and rapidity of its transformations.[81]

Now, if in the universe there exists only matter and energy, and if matter is merely passive, we must conclude, Tait and Stewart argue, that all physical changes, including the thoughts and actions of living things, are transformations of energy. Accordingly, the following question is 'of the very utmost importance': '*Are all forms of energy equally susceptible of transformation?*'[82] For if there exists some grade of energy which is less capable of transformation, after successive transformations, while the quantity of energy in the system remains the same, the majority will have degraded into the form which is least susceptible of transformation and will be unusable. Heat is the least transformable form of energy and unless a temperature gradient exists no work can be obtained from it. So while energy may be present in a system in the form of heat, none may be available for transformation.

The process of transforming heat into work takes place in the thermodynamic operations of Carnot's perfect heat engine.[83] Tait and Stewart explain that the whole purpose of such an engine – which operates on the reversible Carnot cycle – is to transform heat into work and that its operations, being reversible, enable the greatest amount of work to be obtained from a given amount of heat. The process involves taking the heat which is not converted into work to the condenser and then reinvesting this heat, together with the heat-equivalent of the work done, back into the boiler.

But even a perfect heat engine cannot transform all of the heat which passes through the system into useful work, for the condition which would enable this efficiency – that the temperature of the condenser is absolute zero – can never be achieved. So at each conversion attempt, only a portion of heat is available for transformation into work, the remainder is degraded. And with each successive transformation the heat becomes more degraded, or more dissipated, and so less and less is available for transformation into work.

The analogy of the heat engine helps Tait and Stewart to explain why the present visible universe will one day come to an end. The sun functions as the furnace in the vast heat engine which we call the present visible universe. It radiates energy: a portion reaches earth, supplying life-giving energy; a larger portion still is released into the universe in the form of heat. The loss of its heat causes the sun to cool. From the analogy of the heat engine we know that the availability of energy in the universe will continue to diminish.

Other events are also taking place. Tait and Stewart propose that one day the planets will be drawn into the sun by something like 'ethereal friction'.[84] They will lose their orbital energy, spin into the sun, and merge into its mass. Upon impact, the power of the cooling sun will temporarily be restored as visible energy is converted into heat. The sun will then resume cooling, until the restoration of the next collision, and so on . . .

Still, within this process is the possibility of the formation of new solar systems, formed from the nebulous dust surrounding some of the new coalesced masses. But the process will not continue indefinitely: in general, the total number of masses is still decreasing. A time will come when the matter of the universe is but a solitary cooling mass. It will exhibit no visible motion and will amount to nothing more than a useless store of energy – heat at a uniform temperature.

Thus, according to Tait and Stewart, a combination of processes will affect the end of the present visible universe – the dissipation and degradation of energy, and the aggregation of masses.

The authors' hypotheses: the existence of an unseen universe

Guided by the Principle of Continuity, Tait and Stewart arrive at the existence of something other than the visible – 'an invisible order of things'.[85] They cannot suppose that only the visible universe exists, for both its beginning and end would constitute breaks of continuity. Instead, they must conclude that 'the visible system is not the whole universe, but only, it may be, a very small part of it'.[86]

Tait and Stewart propose that the visible, both in matter and energy, evolved out of an unseen universe and that into this unseen universe it will ultimately retreat. Though the two universes are currently independent they remain 'intimately connected', with exchanges of energy taking place between them, facilitated by a less-than-perfectly-transparent ether.[87] The energy that is stored in the unseen is held for the purposes of new creation. This conception of the unseen provides a novel way of coming to terms with the wasteful loss of the sun's energy through dissipation: the law of conservation of energy applies to the whole system.

The scientific theory which is put forward to explain how the visible evolved out of the invisible is that of William Thomson's vortex atom. The theory, which was communicated to the Royal Society of Edinburgh in February 1867, was the latest speculation on the theory of matter. It regarded the universe's primordial atoms as vortices developed from a pre-existing perfect fluid filling all space. The idea that motion could be a basis for matter was a well-established one, with the first real contribution being the theory of vortex motion suggested by the German physicist, Hermann von Helmholtz (1821–94).[88]

Tait and Stewart's only real objection to Thomson's vortex atom theory was that Thomson had assumed a perfect fluid. In a perfect fluid there is no viscosity and in the absence of viscosity there can be no rotation without outside influence. Therefore, a perfect fluid necessitates a creative act in time which constitutes a break in continuity. Hence, in order to have both unbroken continuity and a vortex atom theory, the invisible universe cannot be a perfect fluid.

In a perfect fluid, as Helmholtz proved, the rotating portions of fluid are arranged in knotted filaments, which once set in motion forever '*maintain their identity*'.[89] In contrast, in a fluid which is not perfect, the permanence of the structures is no longer guaranteed. Our authors saw in this evidence of the non-permanence of the visible order of things.

Thomson's theory impressed and inspired Tait and Stewart. They claimed that Thomson had 'gone further than any one else' in accounting for the origin of life.[90] Indeed, during the 1870s Tait began to teach Thomson's vortex atom theory in his Edinburgh lectures. In his address to the British Association in 1871, Tait shared his hopes for the promising theory: 'Our President's [Thomson's] splendid suggestion of Vortex-atoms, if it be correct, will enable us thoroughly to understand matter, and mathematically to investigate all its properties.'[91] Tait was to invest much of himself in these mathematical investigations during the latter years of his career. He was chasing a full classification of the forms of knotted vortex rings, believing in the existence of a unique form of vortex ring for each of the elements. Eventually, Tait admitted, in *Properties of Matter* (1885), that 'the discovery of the ultimate nature of matter is probably beyond the range of human intelligence'.[92]

Pursuing the chain of continuity in a backwards direction, our authors are led to the possibility of an endless number of invisible universes. In form, their model (Figure 11.5) resembles the Ptolemaic model. Each universe is represented by one of the concentric circles in the diagram.

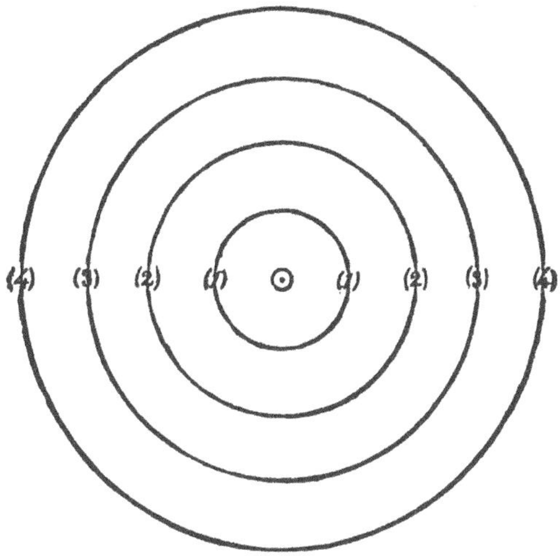

Figure 11.5 Tait and Stewart's concentric model of the Great Whole. From B. Stewart and P. Tait, *The Unseen Universe; Or, Physical Speculations on a Future State*, 2nd edn. London: Macmillan and Co.; 1875: p. 171.

At the centre of the model is an evanescent smoke-ring; produced here on earth, by a smoke box for instance, as in Tait's experiments in 1867. Tait's experiments with smoke rings were conducted for the benefit of William Thomson in order to verify Helmholtz's claims regarding the interaction of vortex rings. The permanence of form of the smoke rings suggested the vortex atom theory to Thomson.

The molecules of this smoke ring – being part of the visible universe – are vortex atoms developed from the unseen. And the entities that constitute the invisible universe are themselves vortex atoms developed from another invisible universe and so on. Pursuing the chain of continuity infinitely far back, Tait and Stewart reach a universe of infinite energy, which has an intelligent developing agency of infinite energy. Together these universes form 'the Great Whole' – a self-contained system, which is 'infinite in energy, and will last from eternity to eternity'.[93]

Superior and angelic intelligences

Our authors do not rely on their multiverse theory, as is often done today with the modern equivalent, as a means of explaining how special our universe is.[94] For them our universe has not been endowed with such fruitfulness on account of the laws of probability but on account of the generosity of the Creator.

Tait and Stewart appreciate the 'delicacy of construction'.[95] In the development of complexity from what is simple: be that the development of compounds from rudimentary elements, or human life from Darwin's primordial germ. In the delicate balance of unstable forces: be that the regularity of the planets, or the abrupt meteorological changes of the sun and earth. And in the uniformity of the construction of atoms: evidence, for Herschel and Clerk Maxwell, that atoms are 'manufactured articles'.[96]

The possibility that other intelligences besides humankind may exist in the universe is not ruled out. Tait and Stewart are open, for example, to the possibility of life on Mars – in agreement with astronomers and physicists of the time, according Tait and Stewart – though they understand that relevant knowledge will not be forthcoming in their own lifetime.

A discussion on the reality of angelic intelligences is not entered into. Tait and Stewart say only that such beings would not belong to the present visible universe; for we cannot perceive them, nor do we imagine that their fate depends on that of the visible universe.

Immortality and the spiritual body

Tait and Stewart do not speculate on the likelihood of personal immortality, having nothing similar to the Principle of Continuity to apply to it, but they do cite two sources in favour of immortality: statements about Christ and man's 'intense longing for immortality'.[97] Whether the latter, being need-driven, constitutes real evidence is questioned by one critic, who regards a materialist accepting this as evidence as 'a creature of the imagination'.[98]

Speculations on immortality are put into three categories of doctrine: (i) the ethereal state, (ii) the bodily state, and (iii) the inconceivability and/or impossibility of a future state. The authors' view is grounded in Scriptural revelation concerning Christ's death and resurrection and the specific nature of His physical transformation. Thus, the nature of our future physical state is bodily but spiritual, or angelic, rather than natural.

Tait and Stewart suppose that immortality is a transference. They propose the following three suppositions:

> It may be regarded as a transference from one grade of being to another in the present visible universe; or secondly, as a transference from the visible universe to some other order of things intimately connected with it; or lastly, we may conceive it to represent a transference from the present visible universe to an order of things entirely unconnected with it.[99]

To these suppositions they apply the Principle of Continuity. The first supposition is discarded on the grounds that the visible universe had its beginning in time and will eventually come to an end. The third supposition is discounted for the following reason. Our authors state two conditions of continued intelligent existence. An individual must: (1) possess an organ of memory, in order to maintain a connection with the past, and (2) be capable of action, or varied movement, in the present. If such an individual, with a connection to the past in one order of things, was to enter an entirely unconnected order of things, they would suffer permanent intellectual confusion that would constitute a breach of continuity. Therefore, only the second supposition remains; hence, immortality must represent a 'transference from the visible universe to some other order of things intimately connected with it'.

Regarding the nature of the transference, the authors' hypotheses follow from their discussion on the latest theories – those of Huxley especially – on the connections between mind and matter, in terms of brain traces and the physical foundations of memory. Tait and Stewart construct a 'frame' for each individual so that they might be connected with the unseen – a 'spiritual body', which receives into it the molecular displacements of the brain.[100] In this conception, the meaning of the anagram in *Nature*, noted earlier, becomes clear.

In the conception of the spiritual body both conditions of continued intelligent existence are fulfilled. The second condition, the capability of action or movement, follows from the author's premise that the unseen is to be full of energy when the visible universe comes to an end. In satisfying both conditions in the conception of the spiritual body, Tait and Stewart feel that they have demonstrated the possibility of the continuance of life beyond the death of the perishable material body.

Tait and Stewart admit that their conception of the spiritual body, as an instrument for personal immortality, may be 'detached from all conceptions regarding the Divine essence'.[101] They maintain, however, that we are logically bound to accept some kind of spiritual body if we accept both the doctrine of immortality and the Principle of Continuity.

Divine action: miracles, the incarnation, and the resurrection

Tait and Stewart believe that in the light of their work there need no longer be a scientific objection to miracles. They are to be regarded as 'transmutations of energy from the one universe into the other'; 'the result of a peculiar action of the invisible upon the visible universe'.[102] The incarnation of Christ presents no problem either: there is no breach of the Principle of Continuity because, in traditional Christian doctrine, Christ submitted Himself to the laws of the universe which are an expression of the Will of the Father. Tait and Stewart suppose that the resurrection of Christ could also have been accomplished without a break in continuity, by an infinite intelligent agency who is capable of developing the visible universe from the unseen.

The authors' practical conclusion

The analogy of Carnot's perfect heat engine inspires the practical conclusion of Tait and Stewart's hypotheses:

> And just as reversibility is the stamp of perfection in the inanimate engine, so a similar reversibility may be the stamp of perfection in the living man. He ought to live for the unseen—to carry into it something which may not be wholly unacceptable. But, in order to enable him to do this, the unseen must also work upon him, and its influences must pervade his spiritual nature. Thus a life *for* the unseen *through* the unseen is to be regarded as the only perfect life.[103]

Reception of *The Unseen Universe*

Early editions of *The Unseen Universe* were widely known and subject to unusually close scrutiny: the authors were successful in securing a readership that would normally have turned away from an inquiry into unseen worlds.[104] Despite being of reduced significance in the scientific world, having incorporated religious doctrine, *The Unseen Universe* made a notable impact on the 'enlightened portion of the public'.[105] Its appeal: the authors' 'real and intimate' knowledge of the latest scientific theories.[106] The dialogue that continued through subsequent responsive editions ensured the longevity of its appeal.

Demand led to a sequel, *Paradoxical Philosophy* (1878), in which the hypotheses of *The Unseen Universe* were reworked into the form of a dialogue conversation between a German mathematician and a select group of the religious and social orthodoxy. It was dedicated to the members of the Paradoxical Society.

Reactions to *The Unseen Universe* were truly diverse. One described the work as: '212 pages of the most hardened and impenitent nonsense that ever called itself "original speculation"'.[107] Another declared: 'nothing as original, and, so far as we judge, satisfactory, whether as regards respect for science, or as giving a logical basis for the belief in man's immortality and the divine rule, has been written this century'.[108] Criticism targeted many areas. (i) The authors' ability to reason philosophically: the book's philosophical sections were weaker than its scientific sections, making for a 'curious compound of severe science and third-rate poetry'.[109] (ii) The legitimacy of their scientific inferences: critics recognized illegitimate and unscientific speculations and cautioned against reliance on analogy.[110] (iii) The effective communication of their hypotheses: the unscientific reader would understand little of the involved science. (iv) Their authority to speak on matters of faith. (v) Issues of religious licence: in order to tie up a philosophical and theological Trinity, Tait and Stewart had produced a Trinity unconnected with the orthodox and 'without warrant from Scripture'.[111] And (vi) their application of science: inappropriately, for the purposes of authenticating the Bible and validating the teachings of Scripture.

Issues with the authors' conception of a spiritual body were numerous and complex, and thoroughly understandable. On a practical level: how is it that spiritual bodies do not interact with one another? Can consciousness exist in two places at once? Does the physical body retire entirely into the spiritual body upon death? . . . The English mathematician and philosopher, Professor William K. Clifford (1845–79), writing in *The Fortnightly Review*, seemed set to demolish the integrity of *The Unseen Universe*, having particular issue with the concept of a spiritual body.[112]

Tait and Stewart dealt robustly with the criticism they received. Later editions were revised and enlarged to make use of criticism and reply to objections; however, they refused to recall any statements. In the preface to the second edition they listed the charges brought against them:

> Some call us infidels, while others represent us as very much too orthodoxly credulous; some call us pantheists, some materialists, others spiritualists. As we cannot belong at once to *all* these varied categories, the presumption is that we belong to none of them. This, by the way, is our own opinion.[113]

Braced for an attack from religious leaders, Tait and Stewart were 'delightfully perplexed', and encouraged, by their response: the two parties agreed on many points and where there were differences of opinion, they were pointed out 'with the utmost courtesy' so as to safeguard the Church's independence and to show due regard for the authors.[114]

Inevitably, Tait and Stewart were accused of treating others badly in the course of their inquiry, particularly theologians, materialists, and spiritualists. It is of little doubt that the work 'wields a very heavy blow at materialism', in the words of one commentator, and that the views expressed about contemporary spiritualists are extremely provocative; however, the claim has no real foundation as regards theologians.[115]

Another challenge faced by the authors was the misrepresentation of their work. Commentators evidently felt qualified, and at liberty, to report on the book's scientific content. They lifted extracts from the book with no regard for context, often inserting terms of their own invention. This careless malpractice infuriated the authors, who addressed the issue publicly in new editions.[116]

Commentators who reacted positively to the book praised its boldness and originality, and applauded the authors' novel ideas which were laid out, without prejudice, in an honest search for the truth and for a noble purpose. Critics recognized an 'earnest religious spirit' that ran through the work.[117] The authors were commended for their applications of 'sound scientific reasoning'.[118] And for their 'many clear expositions of scientific truth'.[119] A number of commentators hoped and believed that there may be some truth in their major propositions.

String theory and M-theory anticipated

In *The Unseen Universe*, Tait and Stewart went far beyond attempting to reconcile science and religion on a few points of conflict: they had proposed a theory of everything; and they were, without question, overambitious in attempting to work out every detail of the unity they saw.

While we cannot remark on the verity of their hypotheses regarding immortality, we might note that some of the scientific hypotheses that they advanced have since reappeared, in modern physics' approaches to a single scientific theory of everything: the concept of an infinite number of universes features in the modern multiverse theory; and Thomson's vortex atom, which was finally abandoned following the acceptance of the theory of relativity around 1910, appears recast in modern string theory.[120]

Having said this, expect to find in the book evidence of imperfections in the contemporary knowledge of science, recognized today as false theories and antiquated notions. Addressing the British Association in 1871, Tait spoke of the value of 'incorrect' scientific hypotheses: 'in science nothing of value can ever be lost; it is certain to become a stepping-stone on the way to further truth'.[121]

Closing remarks

Tait and Stewart – men of both scientific perception and religious instinct – maintained that science and religion are complementary aspects of the same truth. They were described by one critic as modern day examples of Newton and Faraday: 'princes in science, and yet humble, believing Christian men'.[122] Still, they recognized as two distinct groups those who study the How of the universe and those who study the Why:

> A division as old as Aristotle separates speculators into two great classes,—those who study the How of the Universe, and those who study the Why. All men of science are embraced in the former of these, all men of religion in the latter. The former regard the Universe as a huge machine, and their object is to study the laws which regulate its working; the latter again speculate about the object of the machine, and what sort of work it is intended to produce.[123]

In our final quotation from the book, Tait and Stewart express their understanding of their roles as men of science, accepting humbly the limits of their own human intellect:

> the position of the scientific man is to clear a space before him from which all mystery shall be driven away, and in which there shall be nothing but matter and certain definite laws which he can comprehend. There are however three great mysteries (a trinity of mysteries) which elude, and will for ever elude his grasp . . . —they are the mystery of matter and energy; the mystery of life; and the mystery of God,—and these three are one.[124]

Notes and references

Please note: Material sourced from Tait's scrapbook, which is preserved by the James Clerk Maxwell Foundation at 14, India Street, Edinburgh, is referred to by the suffix '(tsb)'.

1. Review of Stewart B. and Tait P., *The Unseen Universe*, *The Freeman*, 21 July 1876. (tsb)
2. B. Stewart and P. Tait, *The Unseen Universe; Or, Physical Speculations on a Future State*, 2nd edn, 1875, Macmillan and Co., p. xi.

3. Stewart and Tait, *The Unseen Universe*, 2nd edn, p. 209.
4. Review of Stewart B. and Tait P., *The Unseen Universe*, *The Educational Reporter*, June 1875. (tsb)
5. Review of Stewart B. and Tait P., *The Unseen Universe*, *The Spiritualist Newspaper*, 28 May 1875. (tsb)
6. B. Stewart and P. Tait, Preface to the fifth edition [*The Unseen Universe*, 5th edn], April 1876. pp. xxiii–xxiv (pxxiii). (tsb)
7. William Rowan Hamilton (1805–65): Andrews Professor of Astronomy at Trinity College Dublin, Astronomer Royal of Ireland and Director of Dunsink Observatory; remembered as the inventor of the calculus of quaternions and for his work on optics. For Tait's obituary tribute to Hamilton, see P. Tait, Sir William Rowan Hamilton, *North British Review*, **45**(89), 1866, 37–74.
8. William Thomson (1824–1907): Professor of Natural Philosophy at Glasgow University (1846–99); remembered, especially, for his work in the areas of thermodynamics, electricity, and submarine telegraphy; knighted in 1866; given the title Baron Kelvin of Largs in 1892. For a contemporary biography of Thomson, see R. Flood, M. McCartney, and A. Whitaker (eds), *Kelvin: Life, Labours and Legacy*, Oxford University Press, 2008.
9. A quotation taken from line 146.
10. Lines 146–9 from Book VI of Homer's *Iliad*. Homer, *The Iliad of Homer: Translated into English Accentuated Hexameters*, trans. J. Herschel, Macmillan and Co., 1866, p. 121.
11. *The Glasgow Herald*, 3 December 1875. (tsb)
12. Review of Stewart B. and Tait P., *The Unseen Universe*. *The Nation*, 27 May 1875. (tsb)
13. To establish priority Tait and Stewart referred to the *Nature* article in the book. See Stewart and Tait, *The Unseen Universe*, 2nd edn, p. 159.
14. *The Edinburgh Academy Register 1824–1914*. Edinburgh: The Edinburgh Academical Club; 1914, p. 120.
15. James Clerk Maxwell (1831–79): considered to be one of the greatest physicists of all time; Professor of Natural Philosophy at Marischal College, Aberdeen (1856–60); Professor of Natural Philosophy at King's College, London (1860–5); Cavendish Professor of Physics at the University of Cambridge (1871–9). For Tait's obituary tribute to Maxwell, see P. Tait, James Clerk Maxwell. *Proceedings of the Royal Society Edinburgh*, **10**(105), 1879–80, 331–9. For a contemporary biography of Maxwell, see R. Flood, M. McCartney, and A. Whitaker (eds), *James Clerk Maxwell: Perspectives on his Life and Work*, Oxford University Press, 2014.
16. Philip Kelland (1808–79): Professor of Mathematics at the University of Edinburgh (1838–79), the first Scot to hold the position. For an obituary tribute to Kelland written by Tait and Prof. George Chrystal, see G. Chrystal and P. Tait, The Rev.

Professor Kelland, *Proceedings of the Royal Society of Edinburgh*, **10**(105), 1879–80, 321–9. Chrystal was Chair of Mathematics at the University of Edinburgh (1879–1911) and previously Regius Professor of Mathematics at St Andrews (1877–9).

17. James David Forbes (1809–68): Professor of Natural Philosophy at the University of Edinburgh (1833–60); Principal of the United College, St Andrews (1860–8); invented the Seismometer in 1842; remembered for his work on glaciers. For a biography of Forbes, in which Tait deals with his scientific work, see J. Shairp, P. Tait, and A. Adams-Reilly, *Life and Letters of James David Forbes*, Macmillan and Co., 1873.

18. Senior Wrangler: the student who placed first amongst graduates taking first class degrees in mathematics in a given year at Cambridge. According to Craik, the first Scot to be Senior Wrangler was Alexander Ellice in 1833. See A. Craik, *Mr Hopkins' Men: Cambridge Reform and British Mathematics in the 19th Century* [e-book], Springer, 2007, p. 45f (no.70). DOI: 10.1007/978-1-84628-791-6. Smith's Prizeman: recipient of the Smith's Prize, an annual £25 prize awarded to two students who had excelled in examinations in mathematics and natural philosophy at Cambridge.

19. In 1908 The Queen's College in Belfast – along with The Queen's Colleges in Cork and Galway, and The Royal University – became The Queen's University of Belfast and The National University of Ireland.

20. With an introduction secured by Andrews, Tait initiated correspondence with Hamilton in 1858. Tait and Hamilton met for the first time in 1859, at the meeting of the British Association in Aberdeen.

21. Margaret Archer Porter (1839–1926) was sister to the Porter brothers, William Archer Porter (1825–1890) and James Porter (1827–1900) who Tait had known at Cambridge. P.G.T. and Margaret married on 13 October 1857, in Shankill, Antrim; together they had six children. James Porter went on to become Master of Peterhouse (1876–1900) and Vice-Chancellor (1881–4).

22. C. Knott, *Life and Scientific Work of Peter Guthrie Tait*, Cambridge University Press, 1911, p. 19.

23. Tait was proposed for fellowship by Philip Kelland in December 1860 and was elected on 7 January 1861. He served the Society as Councillor (1861–4), Secretary to the Ordinary Meetings (1864–79), and General Secretary (1879–1901).

24. Freddie died at Koodoosberg Drift, leading a reconnaissance mission under General Macdonald: he was shot through the heart, while making a 50-yard advance on the Boer position. He is remembered as Scotland's amateur golf champion (1896, 1898) and runner-up (1899). See J. Low, *F. G. Tait: A Record, Being his Life, Letters and Golfing Diary*, J. Nisbet, n.d. [1900].

25. The *Challenger* Expedition was a four-year voyage of scientific discovery of the oceans, conducted between 1872 and 1876. Tait was asked by the scientific leader of the expedition, Sir Wyville Thomson (1830–82) to determine

the corrections that would need to be made to the self-recording deep-sea thermometers used during the expedition in order to compensate for the high-pressure conditions.

26. Tait received honorary degrees from the Universities of Ireland (ScD, 1875), Glasgow (LLD, 1901), and Edinburgh (LLD, 1901), and honorary fellowships of the Edinburgh Mathematical Society (1883) and Peterhouse College, Cambridge (1885). He was an honorary member of the academies of Denmark, Holland, Sweden, and Ireland. He received the Keith Prize (1867–9, 1871–3) and the Gunning Victoria Jubilee Prize (1887–90) from the Royal Society of Edinburgh, and the Royal Medal from the Royal Society of London (1886).

27. During his career Tait published 365 papers and 22 books. For his published papers, consult the *Proceedings* and *Transactions* of the Royal Society of Edinburgh and his *Scientific Papers*, published in two volumes by Cambridge University Press (1898 and 1900). For a detailed bibliography, see C. Pritchard, Provisional bibliography of Peter Guthrie Tait. A paper presented at the Peter Guthrie Tait (1831–1901): Centenary Meeting, July 2001, The Royal Society of Edinburgh. <http://www.maths.ed.ac.uk/~aar/knots/taitbib.htm.> Tait's principal publications include: *Dynamics of a Particle* (1856) with Steele, *Sketch of Elementary Dynamics* (1863) with Thomson, *Treatise on Natural Philosophy* (1867) with Thomson, *Elementary Dynamics* (1867) with Thomson, *Elementary Treatise on Quaternions* (1867), *Sketch of Thermodynamics* (1868), *Elements of Natural Philosophy* (1873) with Thomson, *Introduction to Quaternions* (1873) with Kelland, *Recent Advances in Physical Science* (1876), *Heat* (1884), *Light* (1884), *Properties of Matter* (1885), *Dynamics* (1895), and *Newton's Law of Motion* (1899). For the *Encyclopaedia Britannica*, he contributed the articles: 'Light', 'Mechanics', 'Quaternions', 'Radiation and Convection', and 'Thermodynamics'. He wrote biographical essays on: Hamilton, Maxwell, Balfour Stewart, Kelland, Forbes, Rankine, Kirchhoff, Stokes, and Listing. He also published papers on atmospheric, meteorological, and astronomical phenomena; graph theory and recreational mathematics; and education, including the Cambridge mathematical Tripos.

28. Knott, *Life*, p. 37.

29. Knott, *Life*, pp. 36–7.

30. R. Flint, The late Professor Tait: An appreciation by Professor Flint, *The Student*. n.d. [1901], pp. 60–2 (p. 62). (tsb). Reproduced in Knott, *Life*, pp. 44–6.

31. Tait is known to have attended St John's Episcopal Church in Edinburgh, where his body is interred in the family grave. Stewart's religious denomination is indicated by baptismal records which reveal that he was baptised at the Tron Church in Edinburgh in 1828. The Scottish Episcopal Church is a Scottish Christian Church in communion with, but historically distinct from, the Church of England; it is a member of a group of Christian Churches called the Anglican Communion and comprises seven historic diocese, each with a bishop. The Presbyterian Church is not part of the Anglican Communion and differs in its administration from the Scottish Episcopal Church, having no bishops.

32. For the results of Tait and Stewart's experiments see *Proceedings of the Royal Society London*, **14**, 1865; and *Proceedings of the Royal Society London*, **21**, 1878.

33. P. Hartog, Stewart, Balfour (1828–87). Rev. Gooday G, in H. Matthew and B. Harrison (eds), Oxford Dictionary of National Biography, Oxford University Press, 2004 [online edn] ed. Lawrence Goldman, January 2008, <http://www.oxforddnb.com/view/article/26463>. Schuster has 1853 for the date Stewart became Forbes' assistant, see A. Schuster, Balfour Stewart: Memoir of the late Professor Balfour Stewart, *Memoirs and Proceedings of the Manchester Literary and Philosophical Society*, **1**(1), 1888, 253–72.

34. Stewart also served as: Secretary to the Government Meteorological Committee (1867); President of the Society for Psychical Research (1885); President of the Manchester Literary and Philosophical Society (1887) and President of the Physical Society (1887). From the University of Edinburgh Stewart received the honorary degree of LLD.

35. Owens College was founded in 1851. In 1880 it became The Victoria University of Manchester and in 2004 it amalgamated with UMIST and became The University of Manchester.

36. Stewart published one paper on mathematics, On a proposition in the theory of numbers. See *Transactions of the Royal Society Edinburgh*, **21**, 1857, 407–9.

37. P. Tait, Obituary notice of Balfour Stewart, *Proceedings of the Royal Society London*, **46**(285), 1889, ix–xi (p. ix).

38. Tait, Obituary notice of Balfour Stewart.

39. Among Stewart's principal publications were the textbooks: *Treatise on Heat* (1866); *Lessons in Elementary Practical Physics* (1870) with W.W. Haldane Gee; *Conservation of Energy* (1872), and *Primer in Physics* (1872). For the ninth edition of the *Encyclopaedia Britannica* he contributed an article, 'Terrestrial magnetism'. He also presented papers on a variety of subjects to the Royal Societies of London and Edinburgh. An extensive summary of his work is given by his former student and successor at Owens College, Arthur Schuster (1851–1934) in Schuster, Balfour Stewart, pp. 253–72.

40. Hartog, Stewart, Balfour (1828–87), <http://www.oxforddnb.com/view/article/26463>.

41. J. Black and G. Chrystal, *The Life of William Robertson Smith*, Adam and Charles Black, 1912, p. 163.

42. Stewart and Tait, *The Unseen Universe*, 2nd edn, p. 185.

43. Matter and immortality. Review of Stewart B and Tait P, *The Unseen Universe*, *The Daily Review*, 7 May 1875. (tsb). Originally *Ueber den Materialismus*, Leipzig: W.S.W.; 1863.

44. Stewart and Tait, *The Unseen Universe*, p. 162.

45. Review of Stewart B and Tait P, *The Unseen Universe*, *The Monthly Packet*, n.d., p. 419. (tsb)

46. Stewart and Tait, *The Unseen Universe*, 2nd edn, p. 210.

47. Review of Stewart B and Tait P, *The Unseen Universe*, *The Manchester Examiner*, 28 April [1875]. (tsb)

48. Address by Professor P. G. Tait, MA, FRSE, President of the [Mathematics and Physics Section]. *Report of the British Association for the Advancement of Science; Held at Edinburgh in August 1871*, John Murray, 1872, pp. 1–8 (p. 7).

49. John Tyndall (1820–93): Professor of Natural Philosophy at The Royal Institution; remembered for his work on glacial movement, light, sound, magnetism, and radiant heat in the context of gases and vapours.

50. van Wyhe, J., John Tyndall (1820–93). *Victorian Web* [web]. 2002. <http://www.victorianweb.org/science/tyndall.htm>.

51. R. MacLeod, The X-Club: A social network of science in late-Victorian England, *Notes and Records Royal Society London*, **24**(2), 1970, 305–22 (p. 311). In addition to Tyndall and Hirst, members of the X-Club included: Joseph Hooker (1817–1911), Thomas Huxley (1825–95), Herbert Spencer (1820–1903), Edward Frankland (1825–99), George Busk (1807–86), John Lubbock (1834–1913), and William Spottiswoode (1825–83). The X-Club met for the first time in November 1864.

52. J. Tyndall, *Address Delivered Before the British Association Assembled at Belfast*, Longmans, Green and Co., 1874, p. 1.

53. Tyndall, *Address*.

54. Barton, for instance, in R. Barton, John Tyndall, Pantheist: A Rereading of the Belfast Address, *Osiris*, **3**, 1987, 111–34.

55. *Oxford English Dictionary*, Oxford University Press. ‘pantheism, n.’. <http://www.oed.com/>.

56. Tyndall, *Address*, pp. 6–7.

57. Tyndall, *Address*, p. 60.

58. Tyndall, *Address*, p. vi.

59. See Black and Chrystal, *The Life of William Robertson Smith*, pp. 162–6.

60. Black and Chrystal, *The Life of William Robertson Smith*, pp. 162–3.

61. Knott, *Life*, p. 236.

62. Stewart and Tait, *The Unseen Universe*, 2nd edn, p. xii.

63. The Principle of Continuity was ‘first enunciated by Sir William Grove in his inaugural address to the British Association at Nottingham’, according to Review of Stewart B and Tait P, *The Unseen Universe*, *The Guardian*, 23 June 1875. (tsb)

64. Review of Stewart B and Tait P, *The Unseen Universe*, *The Monthly Packet*, n.d. p. 422. (tsb)

65. Review of Stewart B and Tait P, *The Unseen Universe*, *The Monthly Packet*.

66. Review of Stewart B and Tait P, *The Unseen Universe*, *The Guardian*.

67. B. Stewart and P. Tait, Preface [*The Unseen Universe*, 4th edn]. n.d. [1876]. pp. xxiii–xxiv (pvi). (tsb)

68. Review of Stewart B and Tait P, *The Unseen Universe*, *The Guardian*.

69. Stewart and Tait, *The Unseen Universe*, 2nd edn, p. 60.

70. Stewart and Tait, *The Unseen Universe*, 2nd edn, p. 47.

71. Stewart and Tait, *The Unseen Universe*, 2nd edn, pp. 131–2.

72. Stewart and Tait, *The Unseen Universe*, 2nd edn, p. 125.

73. Stewart and Tait, *The Unseen Universe*, 2nd edn, p. 169.

74. Stewart and Tait, *The Unseen Universe*, 2nd edn, p. 47.

75. Abiogenesis: the hypothesis that living matter can be produced from non-living matter; a term used first by T.H. Huxley in 1870. Spontaneous generation: 'the development of living organisms without the agency of pre-existing living matter, usually considered as resulting from changes taking place in some inorganic substance'. Definitions from: *Oxford English Dictionary*. Oxford University Press. 'abiogenesis, n.', 'spontaneous, adj.'. <http://www.oed.com/>.

76. Stewart and Tait, *The Unseen Universe*, 2nd edn, p. 179.

77. Stewart and Tait, *The Unseen Universe*, 2nd edn, pp. 185–6.

78. Review of Stewart B and Tait P, *The Unseen Universe*, *The Monthly Packet*.

79. Stewart and Tait, *The Unseen Universe*, 2nd edn, p. 72.

80. Stewart and Tait, *The Unseen Universe*, 2nd edn, p. 73.

81. Stewart and Tait, *The Unseen Universe*, 2nd edn, p. 81.

82. Stewart and Tait, *The Unseen Universe*, 2nd edn, p. 82.

83. Carnot's perfect heat engine was a concept developed by the French engineer, Sadi Nicolas Léonard Carnot (1796–1832) in *Réflexions sur la puissance motrice du feu et sur les machines propres à développer cette puissance* (1824), a book on the thermodynamic workings of the steam engine. S.N.L. Carnot was the eldest son of Lazare Nicolas Marguérite Carnot (1753–1823), author of *Géométrie de position* (1803).

84. Stewart and Tait, *The Unseen Universe*, 2nd edn, p. 91.

85. Stewart and Tait, *The Unseen Universe*, 2nd edn, p. 157.

86. Stewart and Tait, *The Unseen Universe*, 2nd edn.

87. Stewart and Tait, *The Unseen Universe*, 2nd edn, p. 158.

88. Hermann von Helmholtz (1821–94): initially he served as a doctor in the Prussian army; then he was, successively, Professor of Physiology at Könisberg, Bonn and Heidelberg; in the late 1860s he made the transition from physiology to physics and in 1871 was appointed to the chair of physics at Berlin. For his paper on vortex motion, 'Über Integrale der hydrodynamischen Gleichungen, welche den Wirbelbewegungen entsprechen', see *Journal für die Reine und Angewandte Mathematik*, **55**, 1858, 25–55. For Tait's translation of the paper into English, see *Philosophical Magazine*, **33**, 1867, 485–512.

89. P. Tait, *Properties of Matter*, Adam and Charles Black, 1885, p. 20.
90. Stewart and Tait, *The Unseen Universe*, 2nd edn, p. 186.
91. Address by Professor P.G. Tait, pp. 1–8 (p. 6).
92. Tait, *Properties of Matter*, p. 15.
93. Stewart and Tait, *The Unseen Universe*, 2nd edn, p. 172.
94. In the multiverse theory there exists a multitude of unobservable universes but, owing to the fine-tuning of natural constants, only one with suitable conditions for our survival. The historical development of the theory is given in H. Kragh, *Higher Speculations: Grand Theories and Failed Revolutions in Physics and Cosmology*, Oxford University Press, 2011.
95. A phrase used repeatedly in the book.
96. Stewart and Tait, *The Unseen Universe*, 2nd edn, p. 167.
97. Stewart and Tait, *The Unseen Universe*, 2nd edn, p. 166.
98. Review of Stewart B and Tait P, *The Unseen Universe*, *The Echo*, 11 June 1875. (tsb)
99. Stewart and Tait, *The Unseen Universe*, 2nd edn, pp. 66–7.
100. Stewart and Tait, *The Unseen Universe*, 2nd edn, p. 159.
101. Stewart and Tait, *The Unseen Universe*, 2nd edn, p. 198.
102. Stewart and Tait, *The Unseen Universe*, 2nd edn, p. 189.
103. Stewart and Tait, *The Unseen Universe*, 2nd edn, p. 192.
104. Review of Stewart B and Tait P, *The Unseen Universe*, *The Manchester Examiner.*
105. Spiritualism and The Unseen Universe. Review of Stewart B and Tait P, *The Unseen Universe*, *The Nottingham Journal*, 16 August 1876. (tsb)
106. Review of Stewart B and Tait P, *The Unseen Universe*, *The Guardian.*
107. Transcendental physics. Review of Stewart B and Tait P, *The Unseen Universe*, *Pall Mall Gazette*, 22 June 1875. (tsb). Quoted by Tait and Stewart in the preface to the second edition, p. vi.
108. Review of Stewart B and Tait P, *The Unseen Universe*, *Edinburgh Courant*, 28 April [1875]. (tsb)
109. J. Hopps, The Unseen Universe; Or Physical Speculations on a Future State: Three lectures by John Page Hopps, *The Truthseeker*, January 1876. p. 3. (tsb)
110. Review of Stewart B and Tait P, *The Unseen Universe*, *The Spectator*, 13 November 1875, p. 1426. (tsb)
111. Review of Stewart B and Tait P, *The Unseen Universe*, *The Christian Treasury*, n.d. [1876], p. 416. (tsb)
112. Clifford's review is reprinted in W. Clifford, The Unseen Universe, in L. Stephen and F. Pollock (eds), *Lectures and Essays*. I. Macmillan and Co., 1879, pp. 228–53. Tait and Stewart's robust response to Clifford's review appears in the preface to their second edition. See Stewart and Tait, *The Unseen Universe*, 2nd edn, pp. viii–x.

113. Stewart and Tait, *The Unseen Universe*, 2nd edn, p. v.

114. Stewart and Tait. Preface [*The Unseen Universe*, 4th edn], pp. xxiii–xxiv (pv). (tsb)

115. Spiritualism and The Unseen Universe. Review of Stewart B and Tait P, *The Unseen Universe*, *The Nottingham Journal*.

116. For instance, in the preface to the 4th edition, Tait and Stewart object to the term 'luminiferous force' which was misquoted in a review of the book which appeared in *The Christian Treasury*. See Stewart B and Tait P. Preface, *The Unseen Universe*, 4th edn, pp. xxiii–xxiv (p. viii). (tsb). See also Review of Stewart B and Tait P, *The Unseen Universe*. *The Christian Treasury*, p. 414. (tsb)

117. Review of Stewart B and Tait P, *The Unseen Universe*, *The Educational Reporter*. (tsb)

118. Review of Stewart B and Tait P, *The Unseen Universe*, *The Inquirer*, 26 June 1875. (tsb)

119. Review of Stewart B and Tait P, *The Unseen Universe*, *The Globe*, 30 April [1875]. (tsb)

120. Knotted strings are at the heart of both Thomson's vortex-atom theory and modern string theory. In modern string theory, vibrating strings are considered to be the fundamental building blocks of matter; the resonant frequencies of the strings determining the type of particle produced. Both theories are discussed in their historical contexts in H. Kragh, *Higher Speculations*.

121. Address by Professor P.G. Tait, pp. 1–8 (p. 5).

122. H. Howat, Events of the past year: Scientific men not undevout, *The Liverpool Mercury*, 27 December 1875. (tsb)

123. Stewart and Tait, *The Unseen Universe*, 2nd edn, p. 2.

124. Stewart and Tait, *The Unseen Universe*, 2nd edn, p. 183.

CHAPTER 12

Faith and *Flatland*

MELANIE BAYLEY

Edwin Abbott's story *Flatland* (1884) seems to be exactly what its name suggests: a tale about quirky geometric creatures living in a flat world. It is well known in the United States as an eccentric little Victorian book, out of the same mould as *Alice's Adventures in Wonderland*. Until recently, though, it was hardly recognized in its native country. It languished on the back shelves of the science libraries, far below the radar of British mathematicians and physicists, let alone students of literature. But *Flatland* has enjoyed a resurgence in recent years. It turns out to be a gold mine for literary critics and scholars with an interest in the crossover between science and fiction.

Flatland encapsulates for modern readers that obscure but all-important link between mathematics and religion that was so fundamental to the Victorians but which has since been lost. Cloaked in a mathematical guise, *Flatland* is a cautionary tale about Victorian dogma and doctrine. It is Abbott's exploration of what he perceived to be the greatest threats posed by science to established religion in the final decades of the nineteenth century. According to *Flatland*, it would be multi-dimensional geometry – not Darwin's theory of natural selection – that would bring down the Established Church if its challenge were not addressed.

The plot of *Flatland*

Flatland's hero and narrator is A. Square: an earnest character and a family man, overly prone to evangelizing. (The square's name is a pun on that of the author, Edwin Abbott Abbott; his parents were first cousins.[1]) He lives in a two-dimensional 'universe' inhabited by other geometric figures: triangles, squares, pentagons, and polygons. His flat society operates according to strictly hierarchical rules where status is determined by the number of an individual's sides. At the bottom of the scale are the isosceles triangles – servants, soldiers, and the police – whose rank is dependent on the angle at their apex, with the armed forces graded the lowest. The middle classes are equilateral triangles; the professional classes are squares and pentagons – our Square is a lawyer – and the 'Nobility' starts with the hexagons and works up, so all polygons of six sides or more are aristocrats.

It is a 'law of nature' that in each successive generation the number of sides should increase by one, so that squares give rise to pentagons and pentagons to hexagons, and so on, so that we eventually arrive at the 'Priestly' class who are circles, or at least many-sided polygons that give the impression of being circles. This feature keeps society in check. Every generation knows that its children will fare better and this gives it hope, so the inhabitants of Flatland live peaceably, never challenging the status quo.

There is one obvious flaw in Flatland's hierarchy; all the women are straight lines. Since intellect is dependent on the angle at a personage's apex and women have no such angle, they are all 'wholly devoid of brain power' according to the Square. They are prone to 'fits of fury' and run the risk of accidentally killing all their families by piercing them through with their sharp points in a particularly violent frenzy. Women are creatures to be pitied, feared, and, above all, contained.

The plot thickens

The role of *Flatland* as caustic satire is not difficult to spot. Although the humour is dry and the narrative voice so relentlessly earnest that the wealth of topical issues it parodies is almost impossible to tease apart, the butt is evidently late Victorian England. Abbott was an advocate of women's rights, not a misogynist. But the story is divided with near-mathematical precision into two halves. While the first is given over to social ridicule, the second adopts a very different tone. It is overtly evangelical and addresses the topic that Abbott described as the primary subject of the text: the notion of faith – religious faith – or faith versus experience.

One day a Sphere arrives in the Square's flat world and tells him about a land of three dimensions. The Square fails to be convinced by the Sphere's mere intellectual argument, so his visitor lifts him bodily out of his world to show him what he means. For the first time the Square sees his 'universe' laid bare to him and realizes its limitations. His mind grasps the notion that lands may exist that he can never fully conceptualize; worlds of three or even four or more dimensions. He comes to realize that reasoned thinking based on experience may not be enough to understand all existence. What happens when self-evident facts are not as concrete as they seem? In a moment of revelation, he grasps the idea that some concepts require more than logical thought; they call for an imaginative leap of faith.

In that notion – in the link between mathematics and leaps of faith – lies the clue to Edwin Abbott's *Flatland*. Masquerading as a fanciful exploration of multi-dimensional geometry, it is a spiritual exercise: a true 'Romance of Many Dimensions' as its author dubs it.

Geometry and religion

An allegorical odyssey through a geometric world seems to us an odd way to explore the scientific challenges facing liberal Christianity in the 1880s, but mathematics presented the perfect metaphor to Abbott. Geometry was linked to Christian beliefs in Abbott's Victorian England in a manner that is completely alien to our way of thinking. The precise reasoning that connects the two is complex and convoluted and will be explored in detail in the second part of this chapter, but Abbott's association between mathematics and religion was entirely logical at the time, given his academic background. The problem was that it relied on some obscure and esoteric connections which were not obvious even to his contemporaries and are yet more arcane now. Put simply, you do not see what you are not looking for, and the layers of meaning that Abbott wrote into *Flatland* sit uneasily in a mathematical fantasy. The language of mathematics is designed specifically to iron out ambiguity and to give precision where everyday speech plays with equivocation. Puns aside, many scientists have little time for multiplicities of meaning even in literary texts, so a geometric fantasy seems the last place to find them.

Just one lone commentator picked up on the story's role as allegory. An anonymous critic, writing in *The Athenaeum*, the leading literary review of its day, declared:

That whimsical book *Flatland* . . . seems to have a purpose, but what that may be it is hard to discover. At first it read as if it were intended to teach young people the elementary

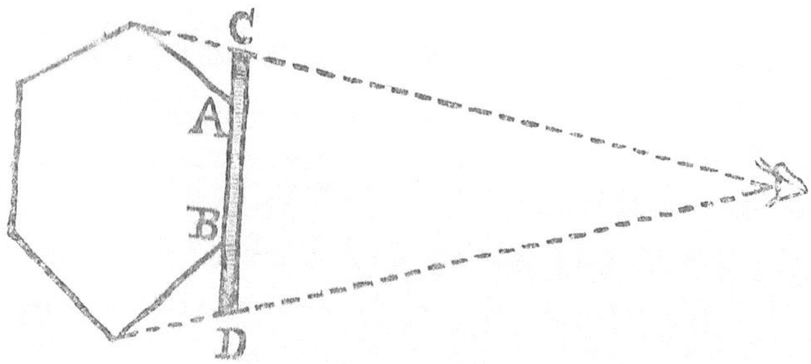

Figure 12.1 The contentious issue of Sight Recognition: Abbott's illustration of how a Hexagon appears to a Flatlander.

principles of geometry. Next it seemed to have been written in support of the more transcendental branches of the same science. Lastly we fancied we could see indications that it was meant to enforce spiritualistic doctrines, with perhaps an admixture of covert satire on various social and political theories.[2]

Uniquely, the *Athenaeum*'s commentator got it right, assuming that his 'spiritualistic doctrines' refer to the more traditional 'theological spiritualism' not to the 'Spiritualism' recently adopted from America describing the practice of communing with the dead.[3] But even he let the 'admixture' drop after that simple passing comment and devoted the rest of the review to mathematical niggles. *Flatland*, he said, was 'spoilt to the mathematical mind by the conception . . . of lines and points as objects which can be seen'. Not surprisingly, when Abbott responded to his critics in his preface to the second edition of 1884, he focused his rebuttals on this perceptual difficulty: of seeing thickness in a two-dimensional world (Figure 12.1).

And so the text's spiritualistic doctrines were forgotten. The multi-layered interpretation that Abbott assumed to be self-evident had passed his readers by, and he finally had to spell out his hidden meaning in a later text: *The Spirit on the Waters* (1897).

Edwin Abbott, the talented teacher

The problem in convincing his readers that *Flatland* was more than a mathematical fantasy was that Abbott had a very real talent for encapsulating complex ideas in simple images. By the time he wrote *Flatland*, he had been headmaster of the City of London School, an independent (private) boys' school at the heart of the capital, for almost 20 years (Figure 12.2). Not surprisingly, he was good at

Figure 12.2 (a) Abbott in 1891 and (b) sometime in the 1910s. Both images, © National Portrait Gallery, London.

explaining things. When he chose to write about multi-dimensional geometry, it looked to his contemporaries as if he was adding to his catalogue of useful textbooks. As the *Athenaeum*'s critic spotted in his first and second readings of *Flatland*, the book seemed simply to be a pedagogical aid to teaching the higher mathematics. There were others that came out around the same time, most notably Charles Howard Hinton's *Scientific Romances*,[4] but Abbott's *Flatland* was by far the best. It was immediately hailed in high quarters. The same month that it was published, the newly appointed Savilian Professor of Geometry at Oxford, J.J. Sylvester, wrote to his friend, England's leading mathematician Arthur Cayley, telling him that he had recommended that the undergraduates in his 'numerous and interesting' class should 'procure Flatland (by Abbott of the City of London School) in order to obtain a general notion of the doctrine of space of *n* dimensions'.[5]

It was perhaps not surprising that Sylvester read *Flatland* as a useful exposition of multi-dimensional geometry. He had used a very similar image, of 'infinitely attenuated book-worms in an infinitely thin piece of paper' to put across the same idea in 1869.[6]

Book-worms, Squares, and topical geometry

Sylvester had been talking to the British Association for the Advancement of Science as president of the mathematical division. He had been trying to convey the radically changing nature of his subject; mathematics had been challenged over recent decades by new ideas arriving from the Continent. His audience was made up of interested members of the general public rather than specialist mathematicians and his speech had been adjusted accordingly. He talked of the 'accidental observation . . . of the Quadrinvariant of a Binary Quartic' as being as much of a shock for the German mathematician Gotthold Eisenstein as meeting 'a Gorilla in the country of the Fantees' or 'a White Polar Bear escaped from the Zoological Gardens [in London]'.

Mathematics in the mid-Victorian era was undergoing what the French would call a *bouleversement* – a total philosophical revolution. The changes were slow, detailed, disparate, and cumulative which makes the moment of transformation impossible to pinpoint.[7] But Sylvester was speaking immediately after an event that turned out to be pivotal in the metamorphosis: the arrival of Bernhard Riemann's non-Euclidean geometries in the 1860s. Sylvester devotes a long footnote to them in the published version of his speech, which appeared in *Nature* in the closing days of the decade:

> In philosophy, as in aesthetic, the highest knowledge comes by faith . . . If an Aristotle, or Descartes, or Kant assures me that he recognises God in the conscience, I accuse my own blindness if I fail to see with him. If Gauss, Cayley, Riemann . . . have an inner assurance of the reality of transcendental space, I strive to bring my faculties of mental vision into accordance with theirs . . . I acknowledge two separate sources of authority—the collective sense of mankind, and the illumination of privileged intellects.[8]

His words have a familiar ring to them; they read almost like a religious credo. Like Abbott in *Flatland*, Sylvester is encapsulating that assumed link between geometry and religion. To understand why he felt the need to speak in such figurative terms to the mathematics section of Britain's leading scientific society, we need to look at the difference between the algebraic analysis and geometry that Sylvester was trying to describe and the mathematics with which the audience would have been familiar.

Victorian mathematics

Broadly speaking, mathematics to the educated Victorian in the street revolved around counting and measuring things. It covered roughly what students would still learn in high school today: arithmetic, simple algebra, and classical geometry – the geometry of triangles and squares and circles. The Victorian approach had been summed up a century before by the renowned mathematician Leonhard Euler, who had declared in his influential textbook on algebra (1770) that 'mathematics is nothing other than the *science of magnitudes*'. The phrase encapsulated the traditional understanding of mathematics so well that in 1847 George Boole (of Boolean logic) echoed Euler's words almost unchanged to describe the mathematics of the day, declaring that it 'is essentially, as well as actually, the Science of Magnitude'.[9] But there was a difference in context between Boole's and Euler's words. Boole was repeating Euler's traditional definition in order to point out its failings. His intention was to establish that mathematics was in the process of becoming a 'Calculus of Logic'. Modern pure mathematics defines itself in terms of logic and the philosophy of language. Boole's understanding of the field was not yet fully in line with this modern interpretation but his exposition marks the early steps in the transition.[10] By the time that *Flatland* appeared, the subject had moved on but it was still far from sharing a frame of reference with modern mathematics.[11]

Geometry as *geo* + *metria*, or measuring the earth

Mathematics lay at the heart of the Victorian education system. It was seen as an essential field of study and much effort went into making it accessible to the man in the street. Not only did Victorian mathematicians write prolifically for working-class readers – one of the century's leading mathematicians, Augustus De Morgan, wrote 850 articles for the *Penny Cyclopedia* alone – the periodical press published widely on mathematical issues.

Traditional mathematics relied heavily on the link between what a student could prove on paper and what he knew instinctively to be true. A circle was best described as the shape you drew if you stuck a pin in a piece of paper and went round it with a pencil tied to a piece of string, not by the equation $x^2 + y^2 = a^2$. George Salmon, the Irish mathematician whose *Treatise on Conic Sections* (1848) remained the standard textbook for a century, summed up the Victorian approach when he wrote: 'We know what a circle is before we know anything about the equation $x^2 + y^2 = a^2$ and any interpretation of this equation differing either by defect or excess from our previous geometrical conception, must be rejected.'[12] In Victorian Britain, space of *n* dimensions simply could not exist; it was not real in the same way as was three-dimensional space. It might be explored by advanced mathematicians like Gauss, Cayley, and Riemann but it was an algebraic anomaly – a 'mathematical pun' as Salmon called it – and did not warrant its own geometry. As we can see from Sylvester's footnote in *Nature*, even leading mathematicians struggled to see it in any other way.

The distinction may seem academic now but it mattered to the Victorians. The intellectual rigour of Greek geometry had traditionally been seen as so fine that Newton had especially reformulated his mechanics to express his theories geometrically, shunning Descartes's algebraic analysis. He called algebra 'the Analysis of the Bunglers in Mathematicks'.[13]

Geometry formed the backbone of an English mathematical education. All schoolboys were required to learn long sections of classical Greek textbooks. A good knowledge of Greek geometry, along with basic algebra, was needed not only to pass university entrance examinations but also to get into the armed forces and the civil service. Modern proofs were shunned and the subject was taught using Euclid's *Elements*, a textbook that had been around for almost two millennia. Purists even worried about using his book in translation, arguing that much of Euclid's subtle reasoning was missed if he was not read in the original Greek.[14]

What made Greek geometry so special to English academics was the style of reasoning that Euclid used. *The Elements* is a classic example of the deductive method. His whole system starts with a very few basic definitions ('A point is that which hath no parts, or which hath no magnitude; a line is length without breadth'), adds to those postulates ('Let it be granted that a straight line may be drawn from any one point to any other point'), and then axioms ('Things which are equal to the same are equal to one another'), and then works through a series of propositions or problems. It has a clear logical sequence – it builds up from simple, apparently incontrovertible statements and gradually develops them into complex arguments.

Greek geometry was seen as a peerless model of reasoning. Euclid's *Elements* was held to be the perfect marriage of absolute truths, rigorous logic, and rational thought, so geometry did not stop at the school gates. University students – particularly those at Cambridge – were required to rote-learn many of Euclid's proofs and reproduce them in exams.[15] Even free-thinking artists like Alfred Tennyson and William Thackeray were required to grind through their Euclids, with limited success; both Tennyson and Thackeray started at Trinity College, Cambridge with high hopes but neither took their degrees.[16] A battered copy of Thackeray's *Elements* survived in his library.[17]

Euclidean geometry and national progress

Mathematics was the focus of a Cambridge education because the university authorities believed that Euclidean geometry, and the Newtonian mechanics that extended it, revealed a special kind of truth. Both Oxford and Cambridge (still essentially the two centres of English learning) held that the purpose of a university was not to produce specialist academics but to provide all undergraduates with a 'liberal education'.[18] The traditional goal of a liberal education had been to cultivate the purest human sentiments, though this had been pared down by the pragmatic Victorians to reining in 'confused thought and bad reasoning'. The mental culture of a young man was best nurtured, it was believed, through studying ancient texts. It mattered little whether the texts addressed the arts or the sciences, as long as they communicated the thoughts of the greatest classical minds. At Oxford, this meant primarily learning the classics; at Cambridge, it meant studying mathematics. Euclid's geometry was taught because it had stood the test of time, not because England needed mathematicians.[19]

Increasingly, though, educators outside Cambridge were viewing England's love affair with Euclid as hindering national progress in mathematics. The modern mathematics that Boole and others were advocating had been embraced on the Continent since the days of the French revolution. Sylvester's presidential address of 1869 had been arguing the case for educational change. He had declared that Euclid's *Elements*, which he described with heavy irony as 'second in sacredness to the Bible alone, and as one of the advanced outposts of the British Constitution',[20] taught the wrong sort of mathematics. Euclid was holding Britain back; England was 'frozen' inside 'mediaeval modes of teaching'. Sylvester told his audience that the mind of the British student should be 'quickened and elevated and his faith awakened by early initiation into the ruling ideas of polarity, continuity, infinity, and familiarisation with the doctrine of the imaginary and inconceivable.' Euclid, he said, should be 'honourably shelved or buried "deeper than did ever plummet sound"'.

Sylvester was quoting from Shakespeare. The line comes from the final act of *The Tempest*, as Prospero relinquishes his magic powers and throws away his magician's accoutrements:

> . . . I'll break my staff
> Bury it certain fathoms in the earth,
> And deeper than ever did plummet sound
> I'll drown my book.[21]

The link with Prospero is not coincidental but is probably due to an association perceived subconsciously by Sylvester but not recognized. It gestures towards a tantalizing link between the Shakespearean character, Euclid's *Elements*, and Abbott's spiritual message in *Flatland*. In his speech, Prospero is talking about his books of spells; Sylvester is talking about Euclid's *Elements*. Although no Victorian, least of all Sylvester, would have seen *The Elements* as a Hogwart-style book of magic, Sylvester is labelling *The Elements* as more than a simple text. Abbott also quotes from Prospero; the magician's words appear in the cloud around the words 'The End of Flatland' (Figure 12.3).

In his illustration, Abbott uses an abbreviated form of the famous lines from Act IV:

> These our actors . . . were all spirits, and
> Are melted into air, into thin air;

Figure 12.3 Prospero's words in Abbott's closing picture at the end of *Flatland*.

And, like the baseless fabric of this vision

. . . shall dissolve. . . .

. . . We are such stuff

As dreams are made on; and our little life

Is rounded with a sleep.

It is always possible that Abbott was directly inspired by Sylvester's speech of 1869 but it seems unlikely. Sylvester and Abbott probably reached independently for *The Tempest* because Prospero's words encapsulated in a familiar form precisely those ideas that both men were struggling to express. *The Tempest* is a Jacobean masque, rich in symbolism, and in Prospero Shakespeare had created a magus – a wise and powerful figure in whom mathematics, transcendental knowledge, and magic merged. Prospero blends the characters and ideas of two of the leading intellectuals of the day – John Dee, who had written the 'Mathematicall praeface' to the English translation of Euclid's *Elements of Geometrie* in 1570, and Francis Bacon, Clerk of the Star Chamber at the court of James I.[22] Abbott was an expert on Bacon, as we shall see later in the chapter, and it is always possible that he knew of the link between Prospero and Dee. It seems unlikely, however, since he mentions nothing of Dee in his writings. Both he and Sylvester probably quoted Prospero's words merely as a useful soundbite; but their use hints at a deeper association. As alumni of St John's College, Cambridge (albeit separated by three centuries) Dee, Abbott, and Sylvester all shared an understanding that transcendental knowledge and Greek geometry were inextricably linked. For Dee in the sixteenth century, this came dangerously close to connecting Euclid's *Elements* with magic – hence Prospero's role as sorcerer. Dee's preface to Euclid's *Elements* defines 'thynges mathematicall' as encompassing both the natural and the supernatural. It left him struggling to throw off accusations of conjuring and sorcery, a dangerous charge with Puritans in the ascendancy. By the time that Sylvester and Abbott were writing, the idea that mathematics was mixed with magic had been lost, along with the office of Witchfinder General. But the lingering concept that geometry represented a special kind of knowledge still echoed through Victorian England. To understand why, we need to look at the philosophy underpinning Euclidean geometry and how the subject came to dominate the curriculum at Cambridge University.

Euclidean geometry as a special kind of truth

For centuries, Euclid's axioms had been seen as representing a special kind of knowledge. A succession of thinkers over the centuries, from John Locke in England to Immanuel Kant in Germany, had argued that Euclid's geometric axioms were unlike other philosophical truths. John Locke had believed that there was a religious lesson to be learned from studying Euclid's *Elements*. He argued that the kind of reasoning involved in understanding a geometric proof – in working systematically through Euclid's *Elements* – was the same as that required to establish secure knowledge of God. To Locke, the study of mathematics was in itself a Godly pursuit.[23]

Immanuel Kant did not share this belief. He saw geometry as a privileged form of reasoning but he did not perceive any religious element to the subject. Kant viewed Euclid's geometric axioms as the principal examples of a priori truths. Geometrical knowledge, he argued, was not like knowledge of the self which, according to Cartesian philosophy, was the only experiential knowledge that could be assumed with certainty. Neither was it derived from experience. It seemed to occupy an intermediate position, spanning the gap between mind and body.[24]

Kant was the philosopher most widely identified with mathematics in Victorian Cambridge, but the influence of Locke still lurked beneath the surface. The leading private coach of the mid-Victorian era, William Hopkins, believed firmly that the higher mathematics had a Christian element. And Hopkins' view was important because, in the fiercely competitive world of mid-nineteenth-century Cambridge, it was Hopkins who spent the crucial tutorial hours with students, not the professors. He tutored all the best mathematicians of the era, including William Thomson, James Clerk Maxwell, and G.G. Stokes. To Hopkins, competitive mathematics was a virtuous pursuit because it revealed the fundamental properties of spatial reality – of God's Creation.[25] When Abbott matriculated at Cambridge, Hopkins was still coaching, although he did not teach Abbott personally. We can visualize Abbott's Cambridge as a kind of melting pot in which ideas drawn from the philosophy of Locke, Kant, and Plato became inextricably fused with Christian theology through the teaching of Euclidean geometry.

Until shortly before Abbott arrived, all Cambridge students, Tennyson and Thackeray included, were required to pass at least part of the Mathematical Tripos in order to proceed to any other degree. By the time Abbott joined, mathematics was optional but any student hoping to compete for a university prize still had to gain second-class honours in order to qualify for the competition. Abbott studied mathematics at Cambridge, ironically not because he enjoyed it but because he excelled at Latin and Greek. He graduated top of his year in classics. Luckily he performed creditably in mathematics too, thanks largely to his training at school (he had attended the City of London School as a boy). He competed for and won the Chancellor's medal for classics in 1861, was made a Fellow of St John's College in 1862, and was ordained as a clergyman in 1863.

Abbott and the new geometries

Abbott's particular education at Cambridge, with its mixture of traditional mathematics and classical literature, gave him unusual insight into the new mathematics when it emerged into the public arena in the 1870s. It allowed him to see beyond the particulars that Sylvester was trying to explain in his presidential speech of 1869, beyond, as Arthur Cayley put it when his turn came to address the British Association in 1883, the 'beautiful detail' with which modern mathematics was 'crowded', to the 'tract of beautiful country' in the distance. Abbott was one of a select handful who grasped the philosophical principles behind Bernhard Riemann's non-Euclidean approach. If he was not a mathematician, or as the Square puts it in *Flatland*, was never one of the '*élite*' practitioners, able to master a 'thorough prosecution of this noble and valuable Art', he had at least received a sufficiently rigorous grounding in the subject to understand its importance. But he believed the relevance of Riemann's new geometries lay not in the details they could add to the crowded mathematical landscape but in the contribution they made to the ongoing debate over the relationship between material proof and religious faith.[26]

The arrival of Riemann's geometry

Riemann's ideas were introduced to the general public by Hermann von Helmholtz, a German physiologist and physicist. For a far-reaching challenge to philosophical precepts, Helmholtz chose a strangely fanciful form in which to present them. In a short, highly imaginative article in *The Academy*, a recently launched topical review, Helmholtz focused on the perceptual difficulties posed by the new geometries rather than their mathematical principles. He illustrated his point using an image familiar from Sylvester's speech: that of creatures living in worlds of two dimensions who could easily find traditional geometry impossible to imagine. Where Riemann's full paper was technical in nature and philosophical in tone and extended the debate about differential analysis in geometry, Helmholtz's was simple – perhaps too simple. By making his message accessible, Helmholtz had inadvertently trivialized Riemann's conclusions: there was too much fantasy and not enough formal discussion for the general public to realize the paper's full import.

Riemann's exposition did not appear in its original form until 1873 when it was finally translated into English by the rising star of English mathematics, W.K. Clifford, and printed in the journal *Nature*. *Nature* was a less technical journal than it is now and Riemann's paper was conspicuously different. It was slotted in between a report on the arrival of a ring-necked parakeet at London Zoo, a letter about the Archduchess Marie Régnier's spaniel, and an article on sun-worshippers. It was a challenge, even for the dedicated reader. In it, Riemann described mathematical space as a continuous path, extendable into infinite dimensions (Riemann asks his reader to think of this extendibility as a 'passing over', hence the notion of 'many-foldedness' or 'manifoldness'). Measurement in Riemann's system depends on the comparison of two relative regions rather than expressing their absolute value: 'the more or less and not the how much', as Riemann puts it.[27]

Clifford knew that Riemann's mathematics would be far too complicated for the public at large to understand so he set about interpreting it more broadly. He gave a series of lectures entitled 'The philosophy of the pure sciences' at the Royal Institution, London, in which he explained the implications of the new geometries: that they deconstructed the notion of Euclid's axioms as transcendental truths and opened up alternative visions of space. The lectures were published in the *Contemporary Review*,[28] a journal that occupied a position something like the review section of today's *Sunday Times* or *New York Times*. What makes them relevant here is that Clifford, like Helmholtz, chose to put across his ideas using a familiar kind of image, of 'people, living in the surface, and having no idea of a third dimension'.

When *Flatland* appeared in 1884, then, it came out at an opportune moment. Stories about creatures living in two-dimensional worlds and working busily on geometries that defied conventional reason were well known to the British public. *Flatland* seemed to be yet another exposition of multi-dimensional geometry – a little fanciful, perhaps, for a mathematical text and its narrator had an oddly evangelical turn of phrase, but that only added to the story's charm. It was not until critics began looking at *Flatland* from a historical perspective that Abbott's primary purpose became clear.[29] Almost uniquely in literary analysis, its author spelled out its role as allegory in a later text, telling scholars he had 'worked out' his conceit 'in a little book called "Flatland" published in 1884'.[30] There were clues in *Flatland*: the story is unlike anything else that Abbott wrote. The other book that he published in 1884, for example, was *The Common Tradition of the Synoptic Gospels in the Text of the Revised Version*, which compares the gospels of Matthew, Mark, and Luke and attempts to trace their common heritage. *Flatland* is so unlike Abbott's other books that it was left out of his original entry in the *Dictionary of National Biography* (1937).

Abbott's literary background

Rosemary Jann has argued in her excellent papers on the subject that *Flatland* really only makes sense if it is read alongside Abbott's theological works.[31, 32] I would extend this idea and argue that we need to consider his literary publications too.

As numerous editors have noted, Abbott was a Shakespeare scholar. He put English literature at the centre of the curriculum at the City of London School and read Shakespeare's plays, along with the works of John Milton and Edmund Spenser, with his older pupils. They in turn reported his 'passionate devotion' to English literature. He wrote two textbooks (relevant here) before the publication of *Flatland*: *A Shakespearian Grammar* in 1877 and *English Lessons for English People* in 1871. The *Grammar* interprets Elizabethan English for Victorian readers and records how informed he was on the subject, but *English Lessons for English People* gives us insight into Abbott's writing in a different way. It contains a chapter on 'Metaphor and simile' that provides us with a wealth of clues for interpreting *Flatland*. He explains that 'A Metaphor is the transference of the relation between one set of objects to another, for the purpose of brief explanation'. Although *Flatland* is allegory, not metaphor, the two ideas are related. *Flatland* takes one idea and uses it to throw light on another; it talks about the perceptual difficulties inherent in multi-dimensional geometry in order to address notions of faith in established religion.

In a section entitled 'Implied Metaphor, the basis of Language' Abbott extends this idea in *English Lessons*. He explains that 'we most naturally describe the relations of those things which are not visible, tangible, etc., by means of those which are visible, tangible, etc.,' and 'This *analogy* is the foundation of all words that express mental and moral qualities'. He then goes on to teach his readers how to tease apart metaphors, and ultimately how to use them correctly. To illustrate his points, he repeatedly uses quotations from Shakespeare, and the great bard does not always come off in the best light. Quoting from *Henry VI* (which is a very bad play and not to be confused with the *Henry IV*s or *Henry V*) Abbott says 'exaggerations like the following must be avoided:

Comets, importing change of times and states,

Brandish your crystal tresses in the sky,

And with them scourge the bad revolting stars.'

He attacks Shakespeare's poor use of personification: 'the liveliest fancy would be tasked to imagine the stars in revolt, and scourged back into obedience by the crystal hair of comets'.

If Abbott feels confident enough to criticize Shakespeare, we can assume that he was comfortable with using Shakespearean imagery. Shakespeare assumed that his audience could, and would, read multiple interpretations into his lines as his actors delivered them. He wrote his plays to work on several different levels simultaneously and Abbott expects his story to do the same.

Flatland: A Romance of Many Dimensions

By the time that Abbott wrote *Flatland*, he had published widely, mainly on theology, but also on sixteenth-century literature and philosophy. He draws on this cultural heritage heavily in *Flatland*. The story is, as its author tells us, 'A Romance of Many Dimensions'. Abbott picks his words carefully here. Not only is there the obvious pun on 'dimensions' and the hint that we should look for something beyond the superficially obvious in the text, but he also chooses to call the novel a 'Romance'.

In literary terms, the word 'romance' has a very different meaning from its everyday connotations. In literature, it means a fantastical tale. The Romantic poets were not so named because they wrote poems about boys meeting girls but because their poetry was fanciful (think of Samuel Coleridge's *Rime of the Ancient Mariner* or John Keats' *La Belle Dame sans Merci*). The definition of romance that critics have usually chosen to apply to *Flatland* is the *Oxford English Dictionary*'s third, 3.a: 'A fictitious narrative, usually in prose, in which the settings or the events depicted are remote from everyday life.'[33] This certainly describes *Flatland* and it seems to be the sense in which that other story published in the early 1880s about multi-dimensional geometry uses the term – Charles Howard Hinton's *Scientific Romances*. (Hinton brought out a short story called 'What is the fourth dimension?' in 1880, which some critics believe may have inspired Abbott to write *Flatland*.[34] The evidence for this is slim, though, and, as we have seen, there are plenty of other sources that Abbott could have drawn on.) There is another sense of 'romance' which is relevant here, an earlier sense, from which this definition number 3 is derived. According to *OED* 1 a romance is: 'A medieval narrative (originally in verse, later also in prose) relating the legendary or extraordinary adventures of some hero of chivalry. Also in extended use, with reference to narratives about important religious figures.' This is the sense in which Abbott is using the term. Although the medieval style of romance is mainly associated with the fourteenth and fifteenth centuries, it was still in use in the late sixteenth century and this was the period on which Abbott was an expert. The obvious example is Edmund Spenser's *The Faerie Queen*, a novel-length poem about a bygone era of knights and quests and chivalry. We should see Abbott's 'Romance' in this medieval sense of the word.

Like *Flatland*, sixteenth-century romances were both topical satire and allegory, but they could also be exercises in piety.[35] The hero underwent a series of trials in which his faith was tested as his virtue was taxed. In Book II of Spenser's *Faerie Queene* for example, Sir Guyon, the Knight of Temperance, comes across two 'naked Damzelles' bathing in a fountain. Spenser's descriptions of the scene are famously voluptuous. He draws the reader in, lavishing temptation upon temptation, demanding that he (implicitly he not she in Elizabethan England) delight in the sensuality of the 'wanton Maidens' as they wrestle in the crystal waves, lifting their 'lilly paps aloft' and revealing their 'amorous sweet spoiles to greedy eyes'. Spenser is inviting the reader to share in Sir Guyon's temptations and so realize how easy it is to stray from the paths of righteousness. Through reading *The Faerie Queene*, a young man will nominally experience a fall from grace without acting on his temptation and so be helped on his own spiritual journey. *Flatland* offers readers the same kind of spiritual journey, not through temptations of the flesh but through titillation of the intellect.

Abbott must have found it difficult to believe that no one except the anonymous writer for *The Athenaeum* spotted what he was trying to do in *Flatland*. With his background in early modern literature, he must have felt that he had made his intentions very obvious. After all, he had spelt them out on the front cover.

There are two versions of the original front cover in general circulation. The original (Figure 12.4), and the more plain of the two versions, which is essentially the same except that the Square's house is missing. We see the same cloud and, along the edges of the page, are drawn the same dimensions – pointland (no dimensions), lineland (one dimension), flatland (two dimensions), and spaceland (three dimensions). Inside the cloud, the other more mysterious dimensions still appear: five, eight, and ten. The only important distinction between the two original frontispieces is that there is a different quotation along the bottom: 'Fie, fie, how frantically I square my talk' on the plain one in place of 'And therefore as a stranger give it welcome' on the more elaborate version.

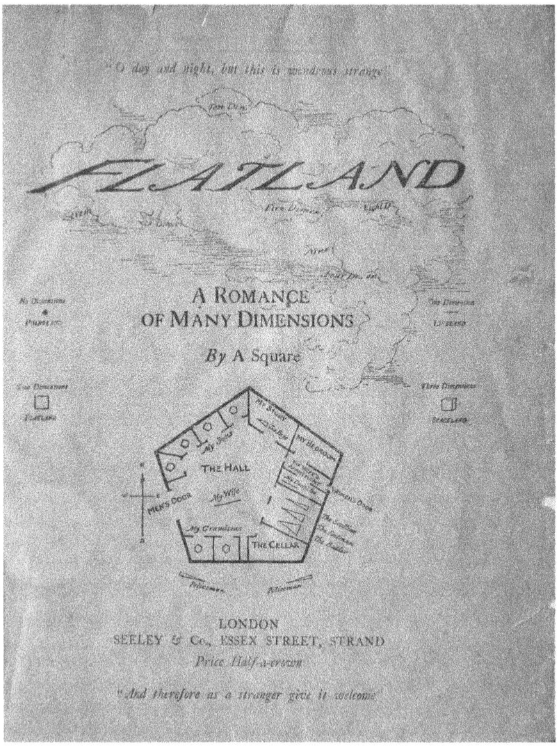

Figure 12.4 Front cover of the first edition of *Flatland*.

The frontispiece cries out to be read as a cryptic message. The cloud dominates both variations, in each case a visual representation of *Flatland*'s fog and obfuscation. Notably, though, the frontispiece is topped and tailed by quotations from Shakespeare, more obviously in the plain version but still prominently in the elaborate. The quotation common to both, 'O day and night, but this is wondrous strange', is from *Hamlet*. It comes from the moment when Hamlet meets his dead father's ghost, in other words when a creature from another dimension comes to earth with an important message for the speaker. That on the plain version 'Fie, fie, how frantically I square my talk' is from *Titus Andronicus*, an early and horribly violent play, and probably has no other relevance than its use of the word 'square'. There could be echoes of other implicit meanings – to 'square my talk' means to adapt or to harmonize what is said in accordance with a preconceived pattern – but reading too much into it involves the sort of mental gymnastics that gives literary criticism a bad name.

The line that replaces it on the second edition is more interesting, probably even to Abbott. In this, the 'Fie, fie' quotation is moved onto the title page and replaced with the subsequent line from *Hamlet*, so the quotations at the top and bottom follow on and reinforce each other. They echo the strange nature of the text.

The two quotations from *Hamlet* are less famous than the lines that follow them: 'There are more things in heaven and earth, Horatio/Than are dreamt of in your philosophy.' The references to 'heaven and earth' and 'philosophy' hint at *Flatland*'s message. Abbott's argument is that, now that geometers have demonstrated that other worlds can exist, Christians have to reassess the basis of their beliefs. Geometers have proved that a fourth dimension may be out there, as real as the three we can see around us. If they are right, and this is not an abstract theoretical concept but a concrete reality, there may be creatures living there that could be in contact with our world, like the Sphere in *Flatland*. Christians have to recognize that any being descending from a fourth dimension would appear as a 'Spirit' to those living in a three-dimensional world. They therefore have to reconsider the fundamental tenets of their belief.

It is not Darwinism that the Church of England has to address if it wants to ward off the threat posed by science, Abbott is saying, but non-Euclidean geometry. To understand his argument better we have to look at the two major works he wrote before and after *Flatland*: *Through Nature to Christ* (1877) and *The Kernel and the Husk* (1886).

'Mr Darwin's theory adapts itself to orthodox religion'

Evolution, Abbott explains in *The Kernel and the Husk*, can be readily accommodated within the existing framework of Christian theology. He says it should be 'welcomed . . . as a luminous commentary on the divine scheme of the Redemption of mankind' because it has thrown new light on 'the unfathomable problems of waste, death and conflict'.[36] It is obvious that Abbott sees evolution as an ameliorating force driven by a divine Will, not as the cut-throat mechanism of neo-Darwinian theory, but that is irrelevant. If Abbott could not grasp fully the nihilistic implications of natural selection, he was in good company. As numerous critics have noted, and Darwin himself admitted, it is far from clear in *Origin of Species* if the author had yet got to grips with the enormity of his theory.[37] In *The Descent of Man* (1871) he confessed that, at the time, 'I was not able to annul the influence of my former belief . . . that each species had been purposely created'.[38]

What is important in Abbott's writing is not that he fails to perceive the bleakness than Darwin himself could not entirely comprehend, but that he genuinely believes that evolutionary theory is compatible with Christian philosophy. In *Through Nature to Christ*, he even claims that 'Mr. Darwin has said of his own theory of evolution, that it adapts itself admirably to Christian teleology'.[39] Notably, though, he does not cite where 'Mr Darwin' said it.

Abbott sees evolution as evidence of the existence of Satan. He devotes an entire chapter in *The Kernel and the Husk* to 'Satan and Evolution':

> I believe that Satan, not God, was the author of the wasteful and continuous conflict that has characterised [evolution . . .] we must learn to think, not of "Evolution by itself," but of "Evolution with Satan." "Evolution without Satan" would appal us by the seeming wastefulness and ubiquity of conflict and the indirectness of its benefits; but "Evolution with Satan" enables us to realize God as our refuge and strength amid the utmost storms and tempests of destruction.[40]

This is more than just a word game for Abbott. The passage reflects his genuine belief that the theory of natural selection can be reconciled with his deep-rooted teleological viewpoint. The world, for Abbott, begins and ends with God. To him, it must reflect design or purpose; it cannot just have come about, as Darwinian natural selection says it must have done, through random chance moulded by competition.

The threat from multi-dimensional geometry

While Darwin's evolution can be accommodated within Abbott's Christian framework, mathematics is more of a problem to him. Now that multi-dimensional geometry has been given a demonstrable existence, Christians must reassess their understanding of God. The defining attributes of a modern God must be more than omniscience, omnipotence, and omnipresence (or, as he summarizes all them in *Flatland*, 'omnividence'). Any being, good or evil, who descends from an unseen fourth dimension could manifest those attributes without being Godly. In the late nineteenth century, Abbott is saying, Christians have to look beyond the scriptures for a modern foundation for their religion.

He sets out his proposed foundation for progressive Anglicanism in *The Spirit on the Waters* (1897), the publication in which he lays bare the central message of *Flatland*. In a chapter entitled, 'How to avoid a wrong conception of God', he warns his readers about the problems involved in conceiving the Divine. He couches his description of man's attempts to understand God in words reminiscent of the 'Feeling' test used by the ignorant and uneducated in *Flatland*: 'When using words about God we are not so much attempting to define God as to feel our way towards communion with him.' Abbott then dismisses traditional notions about God which he labels as 'half-truths, or half-falsehoods':

> What are to be called His supreme attributes? Some would think, and some even venture to say, omniscience, omnipresence, and omnipotence. *This is the broad path that leads to idol-worship.* It conducts us straight to a non-moral or possibly immoral being that has *no affinity with the Christian God* (Abbott's italics).[41]

To show why this must be, he introduces our familiar 'Illustration from Four Dimensions', in which he recapitulates the arguments he left unstated in the second half of *Flatland*:

> We live in a world of three dimensions and find it hard to conceive of a world of four. But let us begin by imagining a world of (practically) two dimensions, in which all the inhabitants are thin Triangles, Squares and Pentagons . . . Next, imagine a living solid looking down from a height on such a Flatland . . . He will see everything that is going on within every house and every Flatland body . . . all will lie open to his view. He will be to them what some of them, perhaps, might be disposed to call the Eye of God. Yet he would not be a God. He would simply be a solid being looking at flat beings, a creature of three dimensions contemplating creatures of two . . . Meantime, what will the Flatlander

see? A Solid he cannot see. But he can see a line of section made by the Solid as it cuts the plane of Flatland; and this line will represent to him a mysterious Being . . . Probably enough he will call it Angel, Ghost, God, or Demon. But it will be simply a common solid creature manifesting itself to flat creatures, a being of three dimensions manifesting itself to beings of two.[42]

He then asks his readers to 'try to conceive the existence of a world of one more dimension than ours', inhabited by 'four-dimensional' or '"super-solid" inhabitants':

A "Super-solid," then, would see what is inside our homes and our bodies. Taking note of the workings of our brains, he might consequently anticipate our thoughts . . . Nowhere in earth or heaven or beneath the earth, could we flee from his presence.

He, then, would be to us what some among us might be disposed to call the All-seeing and Omnipresent God.

But no Christian ought to be able—it is perhaps too much to say "*is* able"—to give the name of God to a Super-solid, who may perhaps be a wholly despicable creature, an escaped convict from the four-dimensional land.

The followers of Christ ought to feel that a good Flatlander is more like God than a bad Super-solid, and that there is nothing essentially divine in being able to see the centres of all the stars in all the solar systems like dots on a sheet of paper.[43]

Abbott had laid out the same argument in *Flatland* in the form of a dialogue between the Square and the Sphere, when the Square first catches sight of his own world from above in Section 18:

Awestruck at the sight of the mysteries of the earth, thus unveiled before my unworthy eye, I said to my Companion, 'Behold, I am become as a God. For the wise men in our country say that to see all things, or as they express it, *omnividence*, is the attribute of God alone.' There was something of scorn in the voice of my Teacher as he made answer: 'Is this so indeed? Then the very pickpockets and cut-throats of my country are to be worshipped by your wise men as being Gods: for there is not one of them that does not see as much as you see now. But trust me, your wise men are wrong.'

I. Then is omnividence the attribute of others beside Gods?

Sphere. I do not know. . . . This omnividence as you call it . . . does it make you more just, more merciful, less selfish, more loving? Not in the least. Then how does it make you more divine?

Mercy and love, or 'moral goodness', are the attributes that Abbott goes on to class as representing the 'highest conception of God' in *Spirit on the Waters*.

Abbott put a similar point forward in *The Kernel and the Husk* (1886) without the reference to *Flatland* when he was arguing the case for a non-miraculous interpretation of Christianity: 'I suppose you will say "A non-miraculous Christ *ought not to be* God to you"? Why not? How does He differ from your conception of God? Is he less loving, less merciful, less just?' The argument here is simple enough, but without the illustration from *Flatland*, the religious discussion of which it forms part becomes pious and protracted. In *Spirit on the Waters*, the illustration from four dimensions engages Abbott's readers, making his argument more palatable. But it also goes beyond simple entertainment: it acts as shorthand. It allows Abbott to echo the mathematical papers of Riemann, Helmholtz, and Clifford. It lets him leave unstated all the challenging themes that they addressed – Kantian notions of absolute truth, of transcendental intuition, even of religious certainty. It will, Abbott hopes, convince his readers that Christianity in the late nineteenth century needs to establish for itself a very different philosophical grounding. And this is where Abbott's argument becomes intriguing for those interested in the wider cultural influences of mathematics.

Abbott's theology as Euclidean proof

Abbott believes that what the Christian faith now needs is a sound axiomatic basis. Although he stresses the limitations of mathematics in the search for God – 'on the one hand it may lead us to vaster views of possible circumstances and existences; on the other hand it may teach us that the conception of such possibilities cannot, by any direct path, bring us closer to God'[44] – his solution to the problem echoes that subliminal bond learned at Cambridge between Euclid's geometry, reasoned argument and liberal Christianity. He confidently sets out five new religious axioms:

1. The things that are seen are temporal, that is not real in the highest sense; the things that are not seen are eternal, that is 'divinely real' . . .
2. There is such a thing as evil, and it is to be regarded as springing from an 'adversary'. . .
3. There is such a thing as forgiveness of sins, and it is a moral and spiritual act . . .
4. . . . God's love and fatherhood are like human love and fatherhood . . .
5. We cannot know God without loving Him.[45]

Ignoring Abbott's theology, which may make modern readers feel edgy, the format that he chooses in *Spirit on the Waters* is uncannily familiar. It mirrors the opening pages of Euclid's *Elements* in which the ancient geometer lays out his axioms:

> I Things which are equal to the same are equal to one another.
>
> . . .
>
> VIII Magnitudes which coincide with one another, that is, which exactly fill the same space, are equal to one another.
>
> IX The whole is greater than its part.
>
> X Two straight lines cannot enclose a space.[46]

There is a fundamental difference, though, between Euclid's geometric axioms and those that Abbott proposes for his 'natural Christianity'.[47] Euclid's axioms were generally held to be self-evident truths. Although their universal applicability had been called into question by the likes of János Bolyai in Hungary, Nikolai Lobachevskii in Russia, and Bernhard Riemann in Germany, they were generally acknowledged as being incontrovertible in a world of three dimensions. Abbott returns to this familiar format in *Spirit on the Waters* at least in part because the geometrical allusion links his axioms to the Kantian notions of absolute truth and transcendental intuition that were generally held to apply to Euclid's propositions. Riemann's paper 'On the hypotheses which lie at the basis of geometry' had questioned the philosophical assumptions of a mathematical system based on Euclid's axioms. Riemann had labelled the axioms as shrouded in philosophical 'darkness'. How much more obscure and philosophically unsound was the statement that 'things that are seen are . . . not real in the highest sense' or that 'there is such a thing as evil, and it is to be regarded as springing from an "adversary" . . . not from God'?

Euclid's reasoning and the religious lexicon

Abbott's decision to use a geometric vehicle for his theological argument must have been a deliberate ploy. Of course, it is always possible that it was an automatic response. Euclidean geometry had been kept at the heart of the Cambridge curriculum specifically because it was the best instrument for educating men in reasoning and Cambridge graduates were so imbued with Euclid's style that they frequently reached for it when faced with a difficult point to argue. Karl Pearson, the biometrician, uses a Euclidean format for his argument when trying to set out his new secular ideology of 'Freethought' (the butt of much of the humour in the first half of *Flatland*) in *The Ethic of Freethought* (1888).[48] But Abbott's choice seems to have been carefully made.

Spirit on the Waters is written more than a decade after *Flatland*, comparatively late in Abbott's career and at a point when he had been preaching and writing on theological issues for more than 20 years. It seems unlikely in the extreme that a man like Abbott, who had devoted his life to examining the steps involved in reasoned or scientific argument and comparing them with the exercise of religious faith, could genuinely have believed that any of his statements could be taken as 'universal truths' as self-evident as the statement that 'The whole is greater than the sum of its parts'. Instead, Abbott seems to have been deliberately mimicking Euclid's style for a purpose.

In the preface to *Spirit on the Waters*, he tells his readers that he has adopted the 'aphoristic' style to 'repel all but those who are genuinely interested'. His decision seems to have the opposite effect, at least to a modern reader, but that matters little if we look at the text's historical context. He classifies 'those who are genuinely interested' as those who find the supernatural elements of the Bible an impediment to Christian belief, in particular the miracles and the 'Miraculous Conception'. By implication, he is directing his argument to those who embrace modern scientific thinking: Comte's Positivism, Darwin's evolutionary biology, and Riemann's geometry. By mimicking Euclid's *Elements*, Abbott deliberately renders *Spirit on the Waters* old-fashioned or anachronistic compared with contemporary theological or even scientific treatises. Euclid's deductive style of reasoning had long been out of date by 1897. Even mathematical texts had been using a more discursive style for decades, and mathematicians had long been trying to throw off the label of deductivism. They argued instead that their discipline should be viewed as an inductive science, one which inferred its general laws and principles from observing the outside world. Sylvester's 'Plea for the mathematician' of 1869 was a counter-argument to Thomas Huxley's claim that maths 'knew nothing of experiment, nothing of induction, nothing of causation'.

Inductive reasoning was important to Abbott. He had been calling on his readers to exercise their powers of induction from his earliest writings.[49] By presenting his theological argument in the style of a deductively-reasoned geometric paper, Abbott reiterates (without stating it) the central theme that runs through much of his theological writing: that, in the closing years of the nineteenth century, Anglicanism needs to embrace the lexicon and reasoning of contemporary science or face the loss of a generation of young scientifically minded Christians. In the language of Darwin, or rather Herbert Spencer, he turns to geometry to give Christianity the fitness that it needs to survive. *The Kernel and the Husk* is written as a series of letters to an unnamed 'doubter' who has suffered a crisis of faith after a single term at university. It takes a strong degree of 'genuine interest' to plough through its full 400 pages. *Spirit on the Waters* is more succinct. While a scientist might argue with the sentiments of the later text, he would find the style more engaging. As Abbott had proved in *Flatland*, evangelical language passes almost unnoticed when delivered in the style of a mathematical treatise.

Abbott's technique in *Spirit on the Waters* is echoed in the closing chapters of *Flatland*, not so much by the Sphere, whose language is studiously secular, but by the Square whose mode of expression is always that of the prophet or preacher. He awakes 'rejoicing' after his episode with the Sphere and decides to 'go forth and evangelize [proclaiming] the Gospel of Three Dimensions'. He talks about 'believers' and 'converts' and 'Truth'. But the Square remains an object of ridicule throughout the story, albeit mitigated by pathos. He slips into hyperbole in his attempts both to prostrate himself before the Sphere and argue for the existence of a fourth dimension: 'Trifle not with me, my Lord' he begs in Section 19:

> your own wisdom has taught me to aspire to One even more great, more beautiful, and more closely approximate to Perfection than yourself. As you yourself, superior to all Flatland forms, combine many Circles in One, so doubtless there is One above you who combines many Spheres in One Supreme Existence, . . . there is yet above us some higher, purer region, whither thou dost surely purpose to lead me—O Thou Whom I shall always call, everywhere and in all Dimensions, my Priest, Philosopher and Friend—some yet more spacious Space, some more dimensionable Dimensionality.

Somehow, in his rational mathematical world, the Square has mastered Biblical phraseology. He uses it unremittingly in his addresses to his 'Priest', the Sphere, so that he sounds halfway between a fictitious Elizabethan courtier and a musical-hall host. As he rambles on, we begin to lose our faith in him. He never quite grasps what Abbott means by the 'One Supreme Existence', which he still sees as a Sphere of Spheres, existing as a physical reality in an invisible dimension – a 'yet more spacious Space'. He never manages to look beyond multi-dimensional reality, to read the primary message of *Flatland*, which we know from *Spirit on the Waters* is to look for God beyond the tangible, beyond dimensionality. Ultimately, the Square is imprisoned intellectually as well as physically.

And this is as it should be. The Romantic model demands that the Square should suffer eternal imprisonment so the reader need not. If the reader chooses to limit himself or herself to that which is written on the page, he or she will share the fate of the Square and stay trapped in a spiritual prison. To experience real spiritual freedom, Abbott is saying, the reader must discover the 'truth beyond the truths'. This lies, he says, in the spiritual world. His final words can be found in a much later text, *Apologia: An Explanation and Defence* (1907): 'As Solidland may be found more "real" than Flatland, so Thoughtland may be found more real than Factland.' What is important to Abbott in *Flatland* is not what is written on the page but on the religious journey that it occasions in the reader.

Notes and references

1. E.A. Abbott, *The Annotated Flatland: A Romance of Many Dimensions*, I. Stewart (ed.), Basic Books, 2002.
2. Our library table. *The Athenaeum*, 1884, 2977. 622.
3. *Oxford English Dictionary Online* (last accessed 6 February 2014).
4. C.H. Hinton, *Scientific Romances*, Swan Sonnenschein, First Series, 1886, Second Series, 1896. C.H. Hinton, *An Episode of Flatland; or How a Plane Folk Discovered the Third Dimension; to which is Added an Outline of the History of Unaea*, Swan Sonnenschein, 1907.
5. K.H. Parshall, *James Joseph Sylvester: Life in Work and Letters*, Clarendon Press, 1998.
6. J.J. Sylvester, Plea for the mathematician, *Nature*, 1, 1869–70, 237–9.
7. R. Flood, A. Rice, and R. Wilson (eds), *Mathematics in Victorian Britain*, Oxford University Press, 2011.
8. Sylvester, Plea for the mathematician.
9. G. Boole, *The Mathematical Analysis of Logic, Being and Essay Towards a Calculus of Deductive Reasoning*, Macmillan, Barclay, & Macmillan, 1847.
10. W. Ewald, *From Kant to Hilbert: A Source Book in The Foundations of Mathematics* 1, Clarendon Press, 1996. T. Gowers, J. Barrow-Green, and I. Leader (eds), *The Princeton Companion to Mathematics*, Princeton University Press, 2008.
11. There is an intriguing overlap between the lives of Boole and Abbott through Boole's wife Mary and his five daughters, Mary Ellen, Margaret, Alicia, Lucy Everest, and Ethel Lilian. Mary Boole lived in London and actively supported women's rights, as did Abbott, and they all shared an interest in multi-dimensional mathematics. As Ian Stewart says in his *Annotated Flatland*, we are left with the feeling that they must have met. Any concrete evidence to verify the hunch is proving elusive, however.
12. G. Salmon, A *Treatise on Higher Plane Curves: Intended as a Sequel to A Treatise on Conic Sections*, Hodges and Smith, 1852.
13. W.G. Hiscock (ed.), *David Gregory, Isaac Newton and their Circle: Extracts from David Gregory's Memoranda 1677–1708*, Oxford University Press, 1937 and

A.R. Hall and M.B. Hall (eds), *Unpublished Scientific Papers of Isaac Newton: A Selection from the Portsmouth Collection in the University Library*, Cambridge University Press, 1962.

14. A. Ellis, *Algebra Identified with Geometry*, C.F. Hodgson & Sons, 1874.
15. J. Richards, *Mathematical Visions: The Pursuit of Geometry in Victorian England*, Academic Press, 1988.
16. *Oxford Dictionary of National Biography Online* (last accessed 2 February 2014).
17. J.H. Stonehouse (ed.), *Reprints of the Catalogues of the Libraries of Charles Dickens and W. M. Thackeray*, Piccadilly Fountain Press, 1935.
18. W. Whewell, Thoughts on the Study of Mathematics, in W. Whewell, *On the Principles of English University Education*, John W. Parker, 1838.
19. Richards, *Mathematical Visions*.
20. Sylvester, Plea for the mathematician.
21. W. Shakespeare, *The Tempest* V, i 55–7.
22. P. Grant, The magic of charity: a background to Prospero, *The Review of English Studies*, 27, 1976, 1–16.
23. J. Richards, God, truth and mathematics in nineteenth century England, in M.J. Nye, J. Richards, and R. Stuewer (eds), *The Invention of Physical Science: Intersections of Mathematics, Theology, and Natural Philosophy since the Seventeenth Century*, Kluwer Academic, 1992.
24. Richards, *Mathematical Visions*.
25. A. Warwick, *Masters of Theory: Cambridge and the Rise of Mathematical Physics*, The University of Chicago Press, 2003.
26. R. Jann, Abbott's *Flatland*: scientific imagination and 'natural Christianity', *Victorian Studies*, 28, 1985, 473–90.
27. B. Riemann, On the hypotheses which lie at the bases of geometry, *Nature*, 8, 1873, 14–17.
28. W.K. Clifford, The philosophy of the pure sciences, i, statement of the question. *Contemporary Review*, 24, 1874. W.K. Clifford, The philosophy of the pure sciences, ii, the postulates of the science of space, *Contemporary Review*, 25, 1875.
29. J. Smith, L. Berkove, and G. Baker, A grammar of dissent: 'Flatland,' Newman, and the theology of probability. *Victorian Studies*, 38, 1996, 129–150.
30. E.A. Abbott, *The Spirit on the Waters: The Evolution of the Divine from the Human*, Macmillan and Co., 1897.
31. Jann, Abbott's *Flatland*.
32. R. Jann, Introduction, in E.A. Abbott, *Flatland*. Oxford University Press, 2006.
33. *Oxford English Dictionary Online* (last accessed 6 February 2014).
34. Jann, Abbott's *Flatland*.
35. H. Cooper, *The English Romance in Time: Transforming Motifs from Geoffrey of Monmouth to the Death of Shakespeare*, Oxford University Press, 2004.

36. E.A. Abbott, *The Kernel and the Husk: Letters on Spiritual Christianity*, Macmillan and Co., 1886.

37. D. Kohn, Darwin's ambiguity: the secularization of biological meaning. *The British Journal for the History of Science*, 22, 1989, 215–39. G. Beer, *Darwin's Plots: Evolutionary Narrative in Darwin, George Eliot and Nineteenth-Century Fiction*, Cambridge University Press, 2000.

38. C. Darwin, *The Descent of Man, and Selection in Relation to Sex* 1, John Murray, 1871.

39. E.A. Abbott, *Through Nature to Christ: Or the Ascent of Worship through Illusion to the Truth*, Macmillan and Co., 1877.

40. Abbott, *The Kernel and the Husk.*

41. Abbott, *The Spirit on the Waters.*

42. Abbott, *The Spirit on the Waters.*

43. Abbott, *The Spirit on the Waters.*

44. Abbott, *The Spirit on the Waters.*

45. Abbott, *The Spirit on the Waters.*

46. R. Simson, *The Elements of Euclid; Viz. The First Six Books, Together With The Eleventh and Twelfth*, J. Collingwood, 1827.

47. Abbott, *The Spirit on the Waters.*

48. K. Pearson, The ethic of freethought, in *The Ethic of Freethought and Other Addresses and Essays*, Adam and Charles Black, 1901.

49. E.A. Abbott, Errors in Reasoning, in E.A. Abbott and J.R. Seeley, *English Lessons for English People*, Seeley, Jackson and Halliday, 1871.

CHAPTER 13

Gödel's 'proof' for the existence of God

C. ANTHONY ANDERSON

Kurt Gödel (1906–78) was the greatest logician/mathematician of the twentieth century, the only figure in formal science comparable in achievement to Albert Einstein (1879–1955) in physical science. In fact, Gödel and Einstein both ended up at the Princeton Institute of Advanced Studies and became close friends. Albert Einstein had fled from Germany, and Kurt Gödel from Austria, to avoid the impending Nazi Terror.

Gödel (Figure 13.1) is best known for his famous Incompleteness Theorems that show that the dream of mathematicians from the time of Euclid cannot be achieved. The proofs are absolutely rigorous and demonstrate that it is impossible to provide axioms for the arithmetic of natural numbers that capture all and only the truths about them. It may be a bit difficult to imagine how such a thing can be proved – but all the experts agree that it has been done. This is by no means all that Gödel proved in and about logic and mathematics, but our topic is his proof of the existence of God. Again the logic of the proof uses the highest available standards of rigour, but, unlike proofs in mathematics and pure logic, it uses premises that are not transparently self-evident. These I will discuss shortly.

Gödel's proof, or better, argument, is a development of the Ontological Argument originally due to St Anselm of Canterbury (*c.* 1033–1109) in the eleventh century (Figure 13.2). If you have taken an introductory course in philosophy or

Figure 13.1 Kurt Gödel. Courtesy of Princeton University Library and the Institute for Advanced Study, Princeton.

a course in the philosophy of religion, you may be familiar with Anselm's argument. It is a fascinating piece of reasoning. However, as soon as you hear it you suspect that it uses some kind of logical trick. I think that suspicion is justified, but it's harder than you might think to show exactly what is wrong.

Here is a terse version of Anselm's argument. Something may actually exist in Reality or it may exist just in the understanding. God is by definition the greatest conceivable being. God exists at least in the understanding since even the fool who says that there is no God understands what he is saying. Suppose God exists *only* in the understanding. Then it would be possible to conceive of a being otherwise like God but existing also in Reality. That would be a conception of a being greater than God, that is we could conceive of a being greater than the greatest conceivable being. That is impossible. Hence our assumption that God exists only in the understanding must be false. Therefore God exists also in Reality. Q.E.D.

Amazing! Brilliant! The theist has defeated the atheist using pure logic! Well, not quite. I'll leave it as a puzzle for you to find the flaw or flaws in this argument. Most of the objections that come immediately to mind do not stand up.[1]

Figure 13.2 Anselm of Canterbury, from a sixteenth-century line engraving. © National Portrait Gallery, London.

But some of them do. Amazingly, some version of this reasoning has appealed to an impressive collection of philosophers. Besides Anselm, there are Rene Descartes (1596–1650), Baruch Spinoza (1632–77), Gottfried von Leibniz (1646–1716), Georg Hegel (1770–1831), and more recently the American philosophers Charles Hartshorne[2] (1897–2000), Norman Malcolm[3] (1911–90), Alvin Plantinga[4] (1932–), and finally our hero Kurt Gödel. Though in 1970 Gödel showed his proof to a colleague at Princeton, he did not publish it; however, it is now available as part of his Collected Works (Figure 13.3).[5] Strictly, what these last philosophers have endorsed is a different argument, also apparently due to Anselm. That argument explicitly employs *modal* notions. What this means will be explained below.

As just explained, Anselm proposed to prove the existence of God essentially from a definition. Rene Descartes revived the argument centuries later. Gottfried Wilhelm von Leibniz observed that the Ontological Argument requires as an implicit premise the claim that *it is possible that God exists*. The term 'God' here is understood as defined as 'the perfect being' or 'the greatest conceivable

Ontological proof
(*1970)

Feb. 10, 1970

$P(\varphi)$ φ is positive (or $\varphi \in P$).

Axiom 1. $P(\varphi).P(\psi) \supset P(\varphi.\psi)$.[1]

Axiom 2. $P(\varphi) \vee P(\sim\varphi)$.[2]

Definition 1. $G(x) \equiv (\varphi)[P(\varphi) \supset \varphi(x)]$ (God)

Definition 2. $\varphi \text{ Ess. } x \equiv (\psi)[\psi(x) \supset N(y)[\varphi(y) \supset \psi(y)]]$. (Essence of x)[3]

$$p \supset_N q \;=\; N(p \supset q). \quad \text{Necessity}$$

Axiom 3.
$$P(\varphi) \supset NP(\varphi)$$
$$\sim P(\varphi) \supset N\sim P(\varphi)$$

because it follows from the nature of the property.[a]

Theorem. $G(x) \supset G \text{ Ess.} x$.

Definition. $E(x) \equiv (\varphi)[\varphi \text{ Ess } x \supset N(\exists x)\,\varphi(x)]$. (necessary Existence)

Axiom 4. $P(E)$.

Theorem. $G(x) \supset N(\exists y)G(y)$,
hence $(\exists x)G(x) \supset N(\exists y)G(y)$;
hence $M(\exists x)G(x) \supset MN(\exists y)G(y)$. ($M$ = possibility)
$M(\exists x)G(x) \supset N(\exists y)G(y)$.

$M(\exists x)G(x)$ means the system of all positive properties is compatible. 2
This is true because of:
Axiom 5. $P(\varphi).\varphi \supset_N \psi :\supset P(\psi)$, which implies

$$\begin{cases} x = x & \text{is positive} \\ x \neq x & \text{is negative.} \end{cases}$$

[1] And for any number of summands.

[2] Exclusive *or*.

[3] Any two essences of x are *necessarily equivalent*.

[a] Gödel numbered two different axioms with the numeral "2". This double numbering was maintained in the printed version found in *Sobel 1987*. We have renumbered here in order to simplify reference to the axioms.

403

404 *Gödel *1970*

But if a system S of positive properties were incompatible, it would mean that the sum property s (which is positive) would be $x \neq x$.

Positive means positive in the moral aesthetic sense (independently of the accidental structure of the world). Only then [are] the axioms true. It may also mean pure "attribution"[4] as opposed to "privation" (or *containing* privation). This interpretation [supports a] simpler proof.

If φ [is] positive then *not*: $(x)N\sim\varphi(x)$. Otherwise: $\varphi(x) \supset_N x \neq x$; hence $x \neq x$ [is] positive, so $x = x$ [is] negative, contrary [to] Axiom 5 or the existence of positive properties.

[4] I.e., the disjunctive normal form in terms of elementary properties[b] contains a member without negation.

[b] Here Gödel uses the abbreviation "prop.", which could be read, in isolation, either as "properties" or "propositions". In the context, however, it is clear that it is properties whose positiveness is under discussion. The related discussion in the excerpts from "Phil XIV" in the appendix, below, explicitly concerns "positive properties". With regard to fn. 4, where the reference to "disjunctive normal form" might lead us to think first of propositions, note that in "Phil XIV", p. 108, Gödel speaks explicitly of properties ("Eigenschaften") that are "members of the conjunctive normal form" of complex properties. An interpretation of fn. 4 is offered in the introductory note, pp. 397–398 above.

Figure 13.3 Gödel's proof as it appears in *Kurt Gödel, Collected Works, Volume III*, edited by Feferman et al. (1995) pp. 403–404, recreated with permission of Oxford University Press, USA.

being'. Leibniz attempted to complete the argument by *proving* that such a being is *possible*. Everyone nowadays agrees that he was not successful in this.

Enter Kurt Gödel – some centuries later. Formal Logic (first conceived, in the beginnings of its modern form, by Leibniz) has come a long way since Anselm, Descartes, and Leibniz. Gödel's version of the Ontological Argument can be expressed and validated in *second-order modal logic*. I will say something about what this means shortly. Gödel did not publish his proof. There is some reason to wonder how seriously he took it. He was certainly religious in some sense, but whether he thought that this proof could contribute to rational religious belief is not quite clear. One colleague at Princeton stated that Gödel's reluctance to publish was because he was worried that people might think 'that he actually believes in God, whereas he is only interested in a logical investigation'.[6]

Second-order logic is that part of symbolic logic that deals with *properties* or *attributes* of things. First-order logic (also known as Predicate Logic or Quantification Theory) contains the means for validating such arguments as 'Churchill is a bald person. Hence, there is at least one bald person.' Second-order logic can validate such reasoning as 'Elizabeth is a royal. Kate is a royal. Therefore,

Elizabeth and Kate have at least one attribute in common.' Of course one doesn't really need formal help in evaluating these two example arguments. But all the more complex reasoning used in mathematics and physics can apparently be treated by the repeated application of these and other logical rules. This is one of the great achievements of modern symbolic logic. It has distilled and then crystallized the fundamental principles of deductive reasoning.

Modal logic treats the logic of such notions as: *necessity*, *possibility*, *impossibility*, *contingency*, and *necessary implication*. This part of logic, although initiated by Aristotle himself and developed some by the medievals, has experienced a spectacular revival since the mid twentieth century. These ideas are very important to philosophy, especially metaphysics. Gödel's Ontological Argument uses a combination of second-order logic and modal logic.

Recall that Leibniz attempted to prove the proposition: *It is possible that God exists*. He defined the term 'God' as that being who has all *perfections*. Then he tried to show that these attributes, perfections, are all compatible with one another. As noted, almost no one nowadays thinks that he succeeded in this task.

Gödel uses, instead of the notion of a perfection, the idea of a *positive attribute*. It follows from the axioms he endorses that the positive attributes are compatible with one another. However, I'm sorry to say, he does not give any very clear explanation of what it means to say that a property or attribute is *positive*. Herein lies one of the problems with his alleged proof. A positive property or attribute is, he says, one that 'contains no negation'. Or it might alternatively be taken, he says, as a sort of aesthetic-ethical value term.

Let us suppose, for the sake of argument that this notion can be made clear, perhaps by examples. Apparently *being omnipotent*, *being omniscient*, and *being perfectly good* are positive properties.

Here are the assumptions, that is, the premises of Gödel's argument, together with various definitions that are used to shorten the formulation of the reasoning.

(A1) A property is positive if and only if its negation is not positive.

If Φ is a property, *being so-and-so*, then its negation (or complement) $\bar{\Phi}$ is *being non-so-and-so*.

(A2) Any property necessarily implied by a positive property is also positive.

Definition: Something is *God-like* if and only if it has every positive property.

(A3) *Being God-like* is a positive property.

(A4) If a property is positive, then *necessarily* having that property is also positive.

Definition: A property is an *essence* of something if it is a property that the thing has that necessarily implies every other property that the thing has.

Definition: A thing *necessarily exists* if any essence of it is necessarily exemplified, that is, if it is necessarily true that some (existing thing) has that essence.

(A5) *Being a necessarily existent thing* is a positive property.

From these five assumptions it follows that:

(G) It is necessarily the case that there is a God-like being.

Probably with other very reasonable assumptions we could further prove that this God-like being is unique, that is that there is only one such.

I will not torture you with the actual details of the proof. You can take my word for it – for now – that there is absolutely no problem about the actual reasoning of the proof. The logical assumptions about necessity and possibility may give one pause, but in the end it appears that the underlying logic is quite sound. For those of you who have some acquaintance with formal modal logic, I note that the modal logic used is S5. This is, almost by universal acclaim, the preferred logical system for modality. If there were anything wrong, it would seem to be a problem with the premises (A1)–(A5).

So there you have it: An alleged proof for the necessary existence of a being having every positive property and provided by the greatest logician since Aristotle. (Well, maybe Gottlob Frege is in the running.)

What shall we say? First, it is very unlikely that anyone who has actually taken the time and effort to understand this proof and its logical underpinnings has been converted to theism as a result of their labours. Still, it would be very interesting indeed if the experts all, or mostly, agreed that here was a watertight, purely logical proof for the crucial belief of an important world-view. Alas (or perhaps, 'Hooray!', if you are an atheist) they do not.

If the argument were to count as a *proof*, taking this in its most definite sense, the premises would have to be self-evident or perhaps just overwhelmingly-plausible-when-carefully-considered. And, as I hinted above, the notion of a *positive* property or attribute is just not clear enough to allow us to reach such a judgement. At least I do not find it such and I am not alone in this. Some would claim further that in the end the argument is *circular*, or in a terminology now being ruined by the media, it *begs the question*. This is a charge that is much easier to make than to sustain. I will not here digress on the topic.

As the argument is actually here formulated, it is subject to a much more serious objection. If we use Gödel's same premises and the same underlying logic, we can prove the following generality:

(P) Any proposition that is true, is *necessarily* true.

Again, I will not here supply the reasoning to show this, but it was first given by J. Howard Sobel[7] (1929–2010). Granted, it is not at all obvious how Gödel's premises yield this conclusion. Again I ask you to just take my word for it until you can check it at your leisure. There is no known flaw in Sobel's reasoning.

I take it that this counts as a *reductio ad absurdum* of Gödel's ontological proof. It seems about as evident as most anything can be that there are true propositions that are *not necessarily* true. Be careful here. The consequence (P) does not say: it is *necessary that* if a proposition is true, then it is true. We can agree with that! It says that every true proposition, say 'Obama won the US presidency in 2008', *could not have been false*. It is absolutely necessary that Obama should have won. This can't be right. It is an absolute necessity that 7 + 5 = 12, but it is not an absolute necessity that there are 50 states in the United States of America. My home state, Texas, might have decided not to join the Union. There are a considerable number of Texans who wish that it had never happened and even now would hope that this deplorable situation changes and Texas becomes its own country. But surely the proposition that there is that particular number of states is contingent, not necessary.

So, by my lights, and pretty much everyone agrees, the argument is defeated. But the story is not over. In my article 'Some Emendations of Gödel's Ontological Argument'[8] I thought I saw where Gödel had gone wrong, weakened some of the premises and altered others. The resulting argument, Gödelian in spirit, is valid and does not have the consequence that every true proposition is necessary. I'll give you the premises and definitions with very little comment.[9]

(A1*) If a property is positive, then its negation is not positive.

This is a weakening of Gödel's (A1).

(A2) Any property necessarily implied by a positive property is also positive.

This is Gödel's second premise.

Definition A property is an *essence** of a thing if it necessarily implies all and only the properties that the thing has necessarily.

Definition Something is *God-like** if and only if it has necessarily all and only the positive properties.

(A3*) *Being God-like** is a positive property.

Definition A thing has *necessary existence** if any essence* of it is necessarily exemplified.

(A4) If a property is positive, then *necessarily* having that property is also positive.

This is the same as Gödel's assumption.

(A5) *Being a necessarily existent* thing* is a positive property.

From these, the exciting conclusion follows (details omitted):

(G*) It is necessarily the case that there is a God-like* being.

In this case we get the bonus that uniqueness follows from the assumptions we have already made. And here we can actually prove by what is called a 'meta-logical' argument that there is no 'modal collapse' that is, these assumptions do not have the consequence that whatever is true is necessarily true.

Well, again, this argument seems (to me) to be seriously weakened by the observation that it is just not clear enough what a positive property really is. It is not that I know how to show that one or more of the assumptions about them are false or probably false. The idea of a positive property is just not definite enough for that either.

Let us then wave a fond farewell to the modal ontological argument. Wait, no, not just yet. I believe that there is a version of the argument, a relative of Gödel's version of the argument, that is valid. That is, the conclusion really follows from the premises. And, it appears, the argument is *cogent*. By this last term I mean that it is reasonable to believe the premises. I will make some qualifications on this latter claim shortly.

To my mind the best version of the modal ontological argument is due to Charles Hartshorne, a famous philosopher noted earlier, associated with a particular worldview called 'Process Philosophy'. He actually published his argument early on but, as far as I know, Gödel's modal ontological argument is independent of Hartshorne's. The latter uses the idea of a *perfect being* and really has only two premises, thus:

(H1) It is possible that there be a perfect being.
(H2) That there is a perfect being necessarily implies that it is *necessarily* the case that there is a perfect being.

About this second premise (H2), it seems we only need to grant that if something perfect exists, then it must have *necessary* existence. A contingently existing being, one that might not have existed, would not be *absolutely* perfect. Using the modal logic mentioned above, namely S5, it follows that there necessarily exists a perfect being. The first premise, (H1), is akin to the assumption that Leibniz tried to prove from his definition of a perfection. And Gödel did prove an analogue of it from his assumptions. Gödel took the term 'God', you will recall, to mean 'the being having all and only positive properties'. Hartshorne's version just takes (H1) as an assumption. We know more or less what it would be for a being to be perfect. It would have to be all-powerful, all knowing, and all-good. And it would have to exist necessarily, if at all. Other attributes may be debatable, but if the argument succeeds we will have shown quite enough.

So the cogency of this argument seems to come down to whether or not it is reasonable to believe that perfection is possible. This doesn't mean that it is possible that some one or other of *us* could become perfect. It means that there is no contradiction and no impossibility implicit in the concept of perfection. A caution is necessary here. Sometimes when we say that such-and-such is possible, we mean something like 'Such-and-such is possible *for all I know*' or 'possible *for all anyone knows*'. A mathematician might say: 'It is possible that Goldbach's Conjecture is true and it is possible that Goldbach's Conjecture is false.' This means 'For all mathematicians know as of the present date'. (Goldbach's Conjecture is that every even number greater than 2 is the sum of two (not necessary distinct) primes.) We might call this *epistemic* necessity. (Where 'epistemic' means 'having to do with knowledge.' Epistemology is that branch of philosophy that is concerned with questions about the scope and limits of our knowledge.) Goldbach's Conjecture, if true, is *necessarily* true in the sense of necessity used in the ontological argument. It *could not have been* false. Some philosophers call the kinds of necessity and possibility in question here *absolute necessity* and *absolute possibility*.

Well and good, but is (H1) reasonable or can it be shown to be reasonable? There are considerations that count considerably in favour of (H1). I will not develop this thought fully, but I will outline the basic ideas. A hypothesis can be confirmed by showing that it explains some facts or other. The possibility of a perfect being can be shown to make sense of Set Theory – the mathematical discipline that many believe can serve as the foundation for absolutely all of mathematics. You may say that Set Theory makes sense already (to mathematicians anyway) and so there is no need for such a hypothesis. I think that this is not quite so. The basic idea is that the unimaginably large collections studied in set theory can be understood in terms of the concepts of an infinite mind and otherwise have not been given a clear meaning. One does not have to postulate the actual existence of such a mind, only the *possibility* of such a mind.

That possibility follows from the possibility of the existence of God. So to that extent the intelligibility of mathematics can be vindicated by means of this latter hypothesis.

Finally, there is some reason to believe that the possibility of a morally perfect (and omniscient) being can be used to justify the objectivity of ethical and moral principles. You may not believe that there are any objective moral rules or principles. If so, this will not count as a reason for you for believing that God is possible. But if you do believe that, say, cruelty is just objectively wrong, period, then *one way* to support this as a moral fact is to appeal to the possibility of the commands or judgements of a morally perfect being. Roughly, the idea is that something is objectively bad if it would be judged bad by a morally perfect being, fully apprised of all the relevant facts of the case. This is not intended as a definition of 'morally bad'. As such, it would be circular.

To sum up: Gödel's modal ontological argument does not seem (to me) to be cogent. Hartshorne's version is more promising. The sole really debatable premise is that it is logically possible that God, a perfect being, exists. This is neither self-evident nor, as far as I can see, provable from self-evident propositions. I have suggested, with very broad strokes, how it might be justified by its fruits – as a hypothesis to allow for a meaningful interpretation of mathematics and as a basis for the objectivity of morality. In the end, this would not produce a *proof* of the existence of God (as Anselm, Gödel, and others apparently hoped). Nevertheless, it would be some considerable reason for believing that there is a God. As is always necessary in such cases, it would have to be balanced against the evidence to the contrary.

Notes and references

1. For a lucid critique of Anselm's Argument, see Alvin Plantinga, *God and Other Minds*, Cornell University Press, 1967.
2. Charles Hartshorne, *The Logic of Perfection*, Open Court, 1962.
3. Norman Malcolm, Anselm's ontological arguments, *Philosophical Review*, **69**, 1960, 41–62.
4. Alvin Plantinga, *The Nature of Necessity*, Clarendon Press, 1974.
5. Kurt Gödel, *Collected Works Volume III: Unpublished Essays and Lectures*, S. Feferman et al. (eds.), Oxford University Press, 1995, pp. 388–404.
6. The colleague was Oskar Morgenstern (1902–77), who recorded this observation in his diary in August 1970 (see Gödel, *Collected Works Volume III*, p. 388).

7. J. Howard Sobel, Gödel's ontological proof, in J. Thomson (ed.), *Being and Saying: Essays for Richard Cartwright*, MIT Press, 1987, pp. 241–61.
8. C. Anthony Anderson, Some emendations of Gödel's ontological proof, *Faith and Philosophy*, 7, 1990, 291–303.
9. A personal note: Most of my published philosophical research has been devoted to logic, especially a branch of the discipline called 'intensional logic'. I wrote and published the Gödel emendation paper because I was coming up for tenure. I was not then a theist, but I felt I needed to show some breadth in my research. Curiously perhaps, or maybe not so curiously, the paper generated more interest than all the rest of my work put together. Papers are still appearing in the philosophical journals offering new emendations and new criticisms of Gödel's ontological argument.

AUTHOR BIOGRAPHIES

C. Anthony Anderson received his PhD from the University of California, Los Angeles, in 1977 under the supervision of Alonzo Church. His dissertation concerned the logic of sense and denotation, an extension of symbolic logic to include reasoning about propositions, attributes, and other concepts. He has held visiting positions in Salzburg and Graz (Austria), in Wuhan (People's Republic of China), and is currently Professor of Philosophy at the University of California, Santa Barbara. During 2010–11 he was Senior Templeton Fellow in Hertford College, Oxford. His main interests and publications are in Intensional Logic, Philosophical Logic, and the Philosophy of Religion.

Melanie Bayley spans the divide between the arts and sciences. She read Natural Sciences at Trinity College, Cambridge before working as a journalist for *Nature*. She studied English Literature at Kingston University and went on to gain a DPhil from Oxford in Victorian mathematics and literature. She is now an Honorary Fellow of Kingston University and has written on maths and fiction for the *New Scientist* and the *New York Times*.

Jean-Pierre Brach is Directeur d'études at the Ecole Pratique des Hautes Etudes (Paris, Sorbonne), where he holds the Chair of 'History of Esoteric Currents in Early Modern and Contemporary Europe', and currently serves as a Board member of ESSWE (European Society for the Study of Western Esotericism). He is the author of *La symbolique des nombres* (Paris, P. U. F., 1994), of an annotated edition and translation of G. Postel's *De admirandis platonicorum numerorum secretis* (Paris, Vrin, 2001), and of numerous articles concerning the history of Christian kabbalah, early modern Alchemy, number symbolism, and nineteenth–twentieth-century French occultism. He is also the editor, with Jean-Pierre Laurant, of *Politica Hermetica*, a yearly publication dealing with the scholarly study of the interactions between esoteric currents and politics. One of the scientific co-editors of the *Dictionary of Gnosis and Western Esotericism* (Brill, 2005), he has also directed (with J. Rousse-Lacordaire) a volume of collected essays entitled *Etudes d'histoire de l'ésotérisme* (Cerf, 2007).

Allan Chapman is a historian of science at Oxford University with particular research interests in the history of astronomy and medicine, and in the relationship between science and the Christian tradition. He has written ten books and co-authored three more, has published numerous academic papers, and does quite a lot of TV and radio work. His latest book is *Stargazers: Copernicus, Galileo, the Telescope and the Church: The Astronomical Renaissance, 1500–1700* (Lion Hudson, 2014).

John Fauvel was Senior Lecturer in Mathematics at the Open University, and a former President of the British Society for the History of Mathematics. A major figure in the history of mathematics scene, he was involved in international studies of the relationships between history and pedagogy of mathematics and was a popular speaker at international conferences and meetings. He was an editor and co-editor of several books, including *The History of Mathematics: A Reader* (with Jeremy Gray). He died in 2001.

Owen Gingerich is Professor Emeritus of Astronomy and History of Science at the Harvard-Smithsonian Center for Astrophysics. His historical publications have ranged from ancient astronomy to the twentieth century, but with a concentration on Copernicus, Tycho, Kepler, and Galileo. His most recent book *God's Planet* (Harvard University Press, 2014), pairs with his earlier *God's Universe.*

Andrew Gregory is Reader in History of Science in the Department of Science and Technology Studies at University College London. He is the author of several books on science in the ancient world, including *Plato's Philosophy of Science*, *Ancient Greek Cosmogony*, and *The Presocratics and the Supernatural*, as well as many articles.

Rob Iliffe is Professor of Intellectual History and the History of Science in the Department of History at the University of Sussex. He has published a number of articles on early modern history and the history of science, and has edited the *Eighteenth Century Biographies of Newton* (Pickering, 2006). He has written the *Very Short Introduction to Newton* (Oxford University Press, 2007), and his *Priest of Nature: The Private Heresy of Isaac Newton* will appear from Oxford University Press in 2015. He is Editorial Director of the online Newton Project, Director of the AHRC Newton Papers Project, and co-editor of the journal *Annals of Science*. His main research interests include: the history of science 1550–1800; the role of science and technology in the 'Rise of the West'; techno-scientific and other roots of the environmental crisis; historical interactions between science and religion; and the life and work of Isaac Newton.

Snezana Lawrence is a Senior Lecturer in Mathematics Education at Bath Spa University. She has been involved with a number of national and international initiatives to promote the use of the history of mathematics in mathematics education. Snezana is on the Advisory Board of the History and Pedagogy of Mathematics Group (an affiliate of the International Commission on Mathematics Instruction), and is Mathematics Programme Director of the Prince's Teaching Institute Schools' Programme. She is on the Editorial Boards of *BSHM Bulletin* (as Associate Editor from January 2013) and *Mathematics Today*. Her website mathsisgoodforyou.com has attracted more than 4 million visitors since its launch in 2005. Snezana is interested in the history and perceptions of mathematical cultures – both in terms of how 'doing mathematics' is seen internally, by various groups of practitioners, and externally, by popular media – and writes and teaches on the links between mathematics, art, architecture, and theism.

Elizabeth F. Lewis has recently completed a PhD in the history of mathematics at the University of St Andrews. She holds a MMath degree in mathematics from St Andrews. Her research interest is the Scottish physicist and mathematician Peter Guthrie Tait (1831–1901). Elizabeth has published papers on Tait's statistical models, Tait's first introduction to complex numbers, and the mathematics of C.-V. Mourey.

Massimo Mazzotti is Associate Professor in the Department of History and Director of the Office for the History of Science and Technology at the University of California, Berkeley. His historical work focuses on three main areas: mathematics and the exact sciences (the technical and social meaning of proving, ordering, standardizing), science and Catholicism (especially after Galileo's trial and late Jesuit science), and technology and society in south-west Europe. Recent books include *The World of Maria Gaetana Agnesi, Mathematician of God* (Baltimore, Johns Hopkins University Press, 2007) and *Knowledge as Social Order: Rethinking the Sociology of Barry Barnes* (Aldershot, Ashgate, 2008).

Mark McCartney is a Senior Lecturer in Mathematics at the University of Ulster. His research and scholarly interests include nonlinear dynamics, mathematical modelling, and the history of science. He is the editor (with Andrew Whitaker) of *Physicists of Ireland* (IOP, 2003) and (with Andrew Whitaker and Raymond Flood) of *Kelvin: Life, Labours and Legacy* (Oxford University Press, 2008) and *James Clerk Maxwell: Perspectives on His Life and Work* (Oxford University Press, 2014).

Mark Richards is an independent researcher and collector with a specialist interest in the life and works of Lewis Carroll. He was Chairman of the Lewis Carroll Society from 1998 to 2013 and is currently executive editor of *The Carrollian*. He has organized Carroll-related events and conferences for over three decades and regularly lectures on diverse aspects of the subject.

Robin Wilson is Emeritus Professor of Pure Mathematics at the Open University, Emeritus Professor of Geometry at Gresham College, London, a former Fellow of Keble College, Oxford, and a former President of the British Society for the History of Mathematics. He has written and edited many books on combinatorics and the history of mathematics, including *Combinatorics: Ancient & Modern* (Oxford University Press, 2013) (with J.J. Watkins).

INDEX

Q

R

S

T

The manufacturer's authorised representative in the EU for product safety is Oxford University Press España S.A. of el Parque Empresarial San Fernando de Henares, Avenida de Castilla, 2 – 28830 Madrid (www.oup.es/en or product.safety@oup.com). OUP España S.A. also acts as importer into Spain of products made by the manufacturer.

www.ingramcontent.com/pod-product-compliance
Ingram Content Group UK Ltd.
Pitfield, Milton Keynes, MK11 3LW, UK
UKHW021135160926
475595UK00020B/75
* 9 7 8 0 1 9 8 7 0 3 0 5 1 *